LOOK
WHO'S
TALKING!

BOOKS BY EMILY HAHN

Hongkong Holiday

England to Me

Purple Passage

The Soong Sisters

China to Me

Romantic Rebels

Animal Gardens

On the Side of the Apes

Once Upon a Pedestal

Lorenzo: D. H. Lawrence and the
Women Who Loved Him

Mabel: A Biography of Mabel
Dodge Luhan

Look Who's Talking!

LOOK WHO'S TALKING!

Emily Hahn

THOMAS Y. CROWELL COMPANY

Established 1834

NEW YORK

To Rose Hahn Dawson, with love

Much of the material in this book appeared originally in *The New Yorker,* in slightly different form.

FIRST EDITION

Designed by Stephanie Krasnow

Library of Congress Cataloging in Publication Data
Hahn, Emily, 1905–
 Look who's talking!
 1. Animal communication. 2. Animals, Habits and behavior of. I. Title.
QL776.H33 1978   591.5′9   78-1625
ISBN 0-690-01659-X

78 79 80 81 10 9 8 7 6 5 4 3 2 1

Contents

Illustrations follow page 54.

Acknowledgments

I should like to thank such patient people as Dr. Lawrence Curtis and his staff, Dr. William B. Lemmon, Charles L. Hanson of the Arizona-Sonora Desert Museum, Dr. Michael Fox, Dr. Philip Ogilvie, and many more for discussing the matter of communication with me and for suggesting other sources. Dr. Gordon Hewes was particularly kind in advising me in research, as was Thomas A. Sebeok. I am indebted to Dr. Duane M. Rumbaugh for his permission to attend the conference of primatologists at Atlanta, and to Janet Groth, who read the typescript and helped me to rearrange it in chapters.

CHAPTER 1

Man Alone

If we choose to let conjecture run wild, then animals, our fellow brethren in pain, disease, suffering and famine—our slaves in the most laborious works, our companions in our amusements—they may partake of our origin in one common ancestor—we may be all melted together.

—*Charles Darwin (aged 28)*

Sometimes I have a dream, or perhaps a nightmare, in which I am all mankind rolled into one, standing in an amphitheater and facing row upon row of all the other lives in the world: animals, birds, and everything else that moves and breathes, even insects and fish. It is horribly quiet, and I feel appallingly alone in the hush. I know and they know that it is a hush that will continue forever, because we can't get through to each other: they can't talk.

Why should I, even in my subconscious, think this so terrible? Why does the thought come out in a nightmare? After all, the world is full of creatures that do talk, and I can't communicate with them either—Russians, Indonesians, Lapps, Zulus. I don't have night-mares about Zulus. On the other hand, I don't habitually *see* Zulus: they are not around me in my daily life, whereas a lot of nonhumans are. And like everyone else, when I was a child I assumed that I could swap thoughts with some of these nonhumans. In this I was merely repeating the pattern of mankind in its infancy—when the world was full of fable, when horses and cats and all the rest of the kingdom talked freely with us. (Ontogeny recapitulates phylogeny.) I remember two occasions when I knew, really knew, that I was communicating with animals, once with a cat and once with a dog.

The cat was a stranger, hanging around in the alley in back of our yard, and I happened to encounter it and made friends with it as children do, against the family's rules because we weren't supposed to have any truck with strange cats. Together this animal and I walked down the alley, and near the end of it we happened on the fresh corpse of a rat someone had thrown out. The cat ate it hungrily while I stood by, and then we resumed our walk together. I knew that the cat thought I had helped it find the rat, and liked me for it. We were companions, and I felt fine about the whole thing.

On the other occasion, nothing so definite occurred, just a feeling I had so strongly that I can still bring it back merely by recollecting. On a warm, quiet night in early summer I sat on our porch steps, my dog beside me, leaning slightly against me as we looked out on the street. Nothing happened. We just sat there, together. I have never since felt any more companionable with any living creature. Nearly all children with pets know this feeling and take it for granted that they can communicate the one with the other, and of course to some extent they do, but it takes a while for a child to accept the fact that his dog or cat will never speak to him in human language, his own.

The first time my small daughter encountered a speaking parrot, she was delighted but not really surprised. We were standing in a pub looking at the bird, which had just said, "Hello," and a Labrador dog belonging to the proprietor entered the room. She asked, without undue excitement, "Can the dog talk too?"

It was hard to explain why a bird can do what dogs or horses can't, and it is harder yet to say just why the bird isn't really talking either. I happen to be acquainted with an antisocial African gray parrot in England named Toto, whose owner brings him out now and then in the evening to show him off to guests. After a few minutes of bad-tempered staring at the company, the bird usually says, "Toto go bye-byes now," and his mistress carries him back to the proper room and puts him into his cage, safe and private under a tea towel. Is Toto really talking? Most people say he isn't, because parrots and other so-called talking birds are merely repeating what they hear. On the other hand, Toto seems to realize that when he makes that particular noise he gets what he wants, removal from the room and the people who offend him. He is rewarded much as a bear is rewarded with a bun when he stands up and begs as the San Diego zoo bus goes by, or a dog gets a cookie when he barks on demand. It is a cause-and-effect arrangement, a bargain fulfilled, a trick.

What, then, is the difference between cause and effect, as in this exchange, and question and answer? When do these interchanges become accepted as language? And what is language, anyway? A lot of words have been expended on this subject, but they do not seem to have led to a totally satisfactory definition. One generalization may be hazarded, however: We humans are very keen to hear talk about animals talking with us—or, to use a less inflammatory phrase, communicating with us. We know, or at least we should know after so many years of observation, that they communicate with each other, not only within their own species but sometimes from one species to another. When a baboon, on guard for his troop, utters a warning bark, animals that are not baboons often take the hint nevertheless and move out of the way of danger. For that matter, dogs often give warning to us in exactly the same fashion. They have other means of communicating as well. Suppose, for example, I am sitting at ease reading the paper, ankles crossed, attention wholly centered on the page until I become aware of a slight discomfort, a moist, cool spot on my top ankle. I look down to see the dog nuzzling my foot. Seeing that he has gained my attention, he trots over to the closed garden door, stands there, and looks back at me over his shoulder. His meaning is unmistakable, but if even then I should fail to get the hint he will bark.

Of course, practically all dogs do this, or something similar. But if one stops to think about it, it is remarkable that our two species should have evolved a code that works so well. It is surprising that a large number of messages can be successfully transmitted between man and animal in these ways. The horse may refuse to move when danger threatens, the cat mews for help for her kittens, in fact, much of our agricultural work rests on this kind of thing, on horses being directed where to go and when to pull or stop, on dogs herding sheep or retrieving birds, on cattle being moved (with the help of the dog) from one pasture to another, even on ferrets that catch rabbits for us, and the cormorants I have seen in China that catch fish for their owners.

If it comes to that, man is not a necessary partner in these exchanges. Animal communicates with animal. In the wild, a hippo's yawn signals to certain birds that he is ready to have his teeth picked, and the birds accept the invitation. We have already seen how the baboon's bark can alert nonbaboons; in similar fashion, birds utter warning calls at sight of predators such as hawks, and other animals are quick to take the hint. There are situations, too, in which mistakes arise, when one species gives a signal that is misunderstood by another. H. Hediger, zoopsychologist and former director of the Zurich zoo, tells of a troublesome affair in another European animal garden where the curator of mammals, pressed for space, put a newly arrived giant red kangaroo into a paddock already occupied by a South American stag. Such a combination is, of course, very bad taxonomically—it might mislead hundreds of schoolchildren and give them strange ideas about geography—but the curator didn't know what else to do for the moment with his Australian kangaroo, and necessity prevailed. After all, he reasoned, the animals were much of a size, and it might work.

For a few minutes it looked as if it might. As long as the kangaroo sat in the typical crouching position of his species at rest, the stag paid him no attention and continued to graze. But after a while the marsupial moved, sat up, and made a hop or two. Immediately the stag attacked, rearing up on his hind legs, thrashing the air with his forefeet, and raining blows on the newcomer, which is the way stags behave when they believe themselves threatened. He did not calm down until the kangaroo, intimidated, dropped again into a crouch and stayed quiet. Mollified, the stag returned to the peaceful occupation of grazing, but he kept a wary eye on the kangaroo, and the minute the latter animal showed signs of hopping he went on the offensive again, flailing and fighting until the unfortunate kangaroo was motionless once more. The sequence was repeated until the curator took the kangaroo out of the enclosure and found him another living space. There was no doubt, he said, that the stag had taken the kangaroo for another stag. Every time the unfortunate beast sat up, his little forelegs dangling, the stag thought he was rearing up threateningly. It was a simple question of faulty communication.

There was another case, somewhat similar, in a zoo where a jenny, or female donkey, was a pet of the director, who divided his affections between Jenny and a Pekingese. Jenny resented the little dog, and one day she went for him, rushing forward with mouth open, teeth bared, and ears laid back. That is the way attacking donkeys behave. But the Peke did not interpret these signals properly, as indications of belligerence. Within his frame of reference, dogs that flatten their ears to their heads as they scamper forward are looking for a game, and he thought Jenny wanted to play. He stood his ground, therefore, his head held high, ears pricked, and tail wagging. Jenny in turn got the wrong message, for when donkeys prick their ears and stand still they are making a friendly greeting; in other words, she interpreted the dog's stance as submissive and her wrath subsided, for animals on the rampage almost never attack submissive beasts. Therefore she merely pretended to attack, bending her head down to nibble slightly and harmlessly on the Peke's back. She walked away afterwards satisfied, honor assuaged, while the dog remained in ignorance of what he had just escaped.

Such incidents, or at any rate our interpretation of them, are absorbing, but to most people they are not enough. Why can't we go farther in our understanding? Surely if we could only know what animals are saying to us, life would be enriched. It seems that in other ways we have reached too many limits. We have gone to the moon; we discuss—surprisingly placidly, considering everything—what might happen if we displace too much ozone; we play with the stars, and turn inward to study our own bodies and psyches until it seems that nothing is left for us to explore. But we still don't know with certainty what goes on inside the brains of the animals who share the world with us, and I am not alone in finding this ignorance frustrating. Interest in the subject goes back a long way, into antiquity, at a time when humanity was even closer to the animal world than it is now.

George Boas, in an article entitled "Theriophily," discusses the philosophical history of love of animals, for which he coined the word. Theriophilists, he explains, have several theories about beasts. Some believe that they are as rational as people, or more rational, or better off without reason. Others think that animals are happier than men, and still others hold that they are more moral than we are. "The whole idea or movement . . . is a reaction against the dogma of the superiority of mankind to all other forms of life," he says, adding that the dogma had two sources, one in pagan antiquity, the other in the Bible. Aristotle expressed the pagan attitude when he said that though man shares his senses and appetites with the animals, he alone has rationality, and the Bible contains a similar idea in Genesis—that God gave man dominion over every other living thing that moves on earth. Many Christians denied that beasts have immortal souls, believing that animal souls, insofar as they exist, are in their blood and die with it. The pagans, however, thought animals do indeed possess souls, and in many of their stories and legends animals play thinking parts, as in the Sanskrit book of parables, the Buddhistic Jataka, Aesop's fables, and others. These tales were told again in the twelfth century and yet again by La Fontaine in the seventeenth.

The first important theriophile, says Boas, was Diogenes of Sinope, who lived several centuries B.C. and is called "the Cynic" from the Greek word for "dog." He sought the purely natural existence, passing over such human savages as were known to his generation because even they lived according to some kind of human law. Because dogs lap water, Diogenes did likewise, declaring cups unnecessary. Because animals don't wear any clothing but their own fur or shells, Diogenes did without all covering but a cloth cloak. Animals took shelter in caves or dens, so Diogenes lived in a wine jar, ate his food raw, and declared that he was free of all man-made laws.

But for all his admiration of animals, Diogenes never claimed that they possess the faculty of reason. It was Plutarch, in the Christian era, who did this, in a dialogue entitled *Gryllus*, in which Odysseus, while shipwrecked on the enchantress Circe's island, has a long talk with one of the animals, formerly a man bewitched by Circe—Gryllus, a pig. Gryllus tells Odysseus that he is perfectly happy as he is. He likes being an animal, because animals are wiser and more prudent than men.

The theme of theriophily was for a time swamped by Christianity, says Boas, because of the Bible and the resulting belief in man's superiority: animals could not possibly rank higher than a man made in God's image and possessed of a soul. However, after sixteen centuries, during which time people began to question such simple theology, theriophily began to make itself known once more. In 1549 there appeared a satirical work called *Circe,* a modernized version of *Gryllus,* in which Ulysses interviews not only the pig but all the animals found on the island. All of them, with one exception, are content to be animals. The exception is the elephant, who in his former life was a philosopher: he alone realizes that humans are better off because they have the faculty of reason. *Circe* was very popular, and variations of the idea have continued almost to our time, says Boas, because "to see society from the point of view of a foreigner—a Persian or Chinese or American Indian—was a favorite device for gaining distance and an apparent objectivity; to see it from the point of view of an animal was even better. . . . The device reappeared in our own time in Edmond Rostand's *Chantecler* (1910)." Also, it might be added, in G. B. Stern's *The Dark Gentleman,* and, of course, George Orwell's *Animal Farm.*

Montaigne was the chief theriophile of the sixteenth century. ". . . In the *Apology for Raimond Sebond II*," says Boas, "he engages in a eulogy of the animals which maintained that their brutal stupidity surpasses all that our divine intelligence can do," pointing out how orderly are the bees, how wise and provident the spider, spinning its web, and so on. If men were to achieve everything animals do, they would have to use reason: It follows, therefore, that animals too can reason. Montaigne gave a number of instances, many of which we could call old wives' tales, of the perspicacity of animals. Added to their native wisdom, they can actually learn skills from us, such as dancing, guiding the blind, and working a treadmill. They can teach us, and indeed have taught us many skills, said Montaigne: weaving (from the spider), building (from the swallow), music (from the birds).

This work affronted many people, attacking as it did their happy belief in man's superiority over animals. Among them, René Descartes was the most formidable critic. Descartes, mathematician and philosopher, he who said, *"Cogito, ergo sum,"* declared that animals are unable to think, being merely machines without soul, mind, or consciousness. They respond to stimuli, but have no power to reason: the animal mind, he said, is like a clock made up of wheels and springs. It follows, therefore, that animals have no language, since speech is so closely related to the power of thought. Not all philosophers accepted this argument. The Englishman Henry More and his compatriot William Cavendish, marquis (later duke) of Newcastle, wrote letters of protest to Descartes, as did the French philosopher Pierre Gessendi, and Descartes felt compelled to explain his position more clearly. He wrote in a letter to one of his critics:

"I don't deny animals even feeling, as long as it depends on bodily organs. My opinion isn't so much cruel to animals as favorable to people." Animal feelings, in other words, are purely somatic, unlike the emotions that stir us through the agency of our thoughts. To the end, however, Descartes did not absolutely recant, and this suited the majority of people, since the concept of animal inferiority was comfortably accepted as a law of nature, which meant the same thing as Divine Will.

Two centuries later, Darwin and his ideas on evolution came as a severe shock, showing as they did that we are indeed all of a piece with the rest of animal life, and more closely related to some species than we might wish. In the midst of so much recasting of thought, it was reassuring to reflect that the other animals were at least outdistanced, definitely and forever, in one regard: they could not talk.

Today most of us are reconciled to Darwin. We are even able to give free rein to our always busy, inquisitive minds and investigate what language can be that we alone have the technique to use it. What is speech? The simplest definition is that it is the most efficient method known of communication. When we come to think of it, it is miraculous, yet so common that we take it for granted. It comprises a large proportion of our infant education; unlike other animals we learn to talk as we learn to walk, so that it seems nearly as basic a part of life as breathing, eating, and sleeping. But there is one important difference, as far as we know: Without precept or example a human infant will learn to walk upright (though the so-called wolf children did not), but speech does not come to the untutored infant so easily or naturally. At least, so it is believed. If you could experiment with that hypothetical baby, what sort of language would the child develop by itself? This is a very old question, still unanswered. At least four men who were powerful enough to carry out the experiment are reputed to have tried: the Pharaoh Psamtik, James IV of Scotland, an Indian emperor whose name I have been unable to discover, and Frederick the Great of Prussia—each of whom in his time arranged for a baby to be isolated in order to discover what tongue the child would speak when it grew old enough to utter words instead of mere babbling noises. King James was confident that the resulting language would be Hebrew, because, he reasoned, that was the tongue in which Adam and Eve communicated with each other. Unfortunately, in each case the baby died before it could prove anything except the fact that infants need company, and without it fade away.

CHAPTER 2

Getting Through

Language is only one form of communication. We ourselves have perfected several codes of facial expressions and gestures: anyone who doubts it has only to watch speakers on a television screen when the sound is off. Other species have developed various methods of communicating with one another. Some depend on sight, some on sound, some on scent, and some on touch. For a long time we could not imagine all these different modes: if an animal did not produce at least a fair approximation of our method of signaling (i.e., speech), we gave up trying to get in touch with it. We could see and hear communication going on in the postures and songs of birds, the silent communion of horses, the playful behavior of puppies, but we stood outside nevertheless. In spite of the fact that we were confident of our superiority, many must have entertained a wistful regret that all this inner life was hidden from us. At least that is one way to explain the eagerness with which the world has always received rumors that some animal has been vouchsafed the gift of speech.

A generation before Descartes published his strictures on reason in animals, the western world was fascinated by a white horse named Morocco that belonged to a Mr. John Banks. Morocco, though not able to talk, understood questions and did practically everything Banks told him to; among the men who marveled at his talents were Shakespeare, Ben Jonson, and Sir Walter Raleigh.

Morocco could dance to music. If somebody rolled the dice and Banks asked Morocco to tell them what the score was, the horse tapped the ground with his foot the right number of times. He could answer yes-no questions the same way, tapping once for yes and twice for no. He answered correctly when asked how old a bystander might be: the man would whisper his age to Banks and, sure enough, Morocco tapped out the right answer. Very seldom did he make a mistake. Finally a contemporary writer, who specialized in magic and juggling tricks and knew how the thing was done, wrote about it. He observed that Morocco, while performing, never took his eyes off his master, Banks, who stood very still until the horse had pawed the ground the right number of times and then, unobtrusively, shrugged one shoulder. Whenever this happened Morocco stopped tapping his hoof.

Milbourne Christopher, a modern writer and magician, has traced the career of Morocco and recounted two incidents that he admits may or may not be true. It was said that Banks took his wonderful animal to France and started giving performances there, but was soon arrested and charged with witchcraft. This happened in Orleans. He was condemned to death by the court, but Morocco saved him by going up to a high church official and kneeling down to him. It was held that if the horse had really been possessed of the devil he could never have gone so close to the holy man, and as a result Banks was declared innocent. According to an equally doubtful story, later, in Rome, Banks and Morocco were both burned at the stake.

Banks's horse may have been the first of his kind, but he was only one of a long line of performing animals, not necessarily equines. Near the end of the eighteenth century, New Yorkers were able to watch a star billed as the Learned Dog, and an equally accomplished pig was rapturously received a few years later. Under the auspices of one Mr. Pinchbeck, the pig could play card games, selecting any one of a number of cards that lay on the floor. It was Pinchbeck himself who exposed the secret method he used to aid the animal in making its selections: whenever the pig's snout hovered over the right card, the man sniffed, making a noise just loud enough for the animal to hear but that was lost on the audience.

Horses, however, seem to have been most popular, for there were many of them. Christopher lists the Learned Little Horse, which entertained the public in Glasgow in 1764; the English Military Horse of Knowledge, which could be seen in 1780; Spottie, who appeared in New York in 1807; and an Australian animal, Mahomet, who could add and subtract as fast as any human: he was first exhibited in 1889. Mahomet's guardian, E. L. Probasco, later gave the game away in an interview with a reporter from London's *Sketch*. He said that his method of communication with the horse had been followed for centuries: the trainers of these animals had always taught them to stamp once for yes and twice for no in the simplest way imaginable, using the position of the trainer's whip. If he tilted the whip's far end toward his shoulder, he said, Mahomet knew that he should stamp out "yes," whereas the whip tilted forward meant "no." But, he added, Mahomet and he had worked together for so long, and the horse was so clever, that he didn't need the whip at all. Mahomet knew what he wanted by watching him for the smallest possible signal, and almost never made a mistake.

It was now time for the appearance of the most famous learned horse of all, Clever (or, in his native German, *Kluge*) Hans. The difference between Hans and his renowned predecessors was that nobody seems to have thought of him as one of the show animals: he was in a class by himself. Also, the people who were astonished by Hans's feats were, like him, of a special class, insulated from the mob. They were intellectuals who might never have heard of Spottie or Mahomet, and it is quite possible that they hadn't. At any rate it seems certain that Hans's owner, a mathematics teacher named Wilhelm von Osten, trained his animal in good faith, and the professors who examined Hans and wrote long papers about him did not link him up with the marvelous horses of Mr. Banks and Mr. Probasco, though much of what Hans did was just like the feats of those other animals. For example, he too tapped his hoof in reply to questions: Mr. von Osten had worked out a complicated code of interpretation of the taps. Hans also had another way to say yes and no: he nodded his head for the one and shook it for the other. But it must be admitted that there are just so many ways a horse can express himself.

It was in 1904 that Hans achieved fame, a propitious year for wonder animals. People were intensely, even emotionally involved in the Darwinian controversy, still trying to prove that human beings are firmly, irrevocably superior to all other forms of life and pointing out that we have souls, while they have not; that we possess the capacity to reason, which they do not; and that we can talk and they cannot. A book by Oskar Pfungst, a psychiatrist, with an introduction by Professor C. Stumpf, both of whom were closely involved in the affair of Clever Hans, was published in America after the dust had settled in 1911: *Clever Hans (The Horse of Mr. von Osten)*. In his introduction Stumpf said that the controversy boiled down to one question: Does an animal possess animal consciousness—the power to think? And if he does, is it similar to human consciousness? As he reminded his readers, the problem is an old one. He gave a quick résumé of Aristotle's ideas on the subject, including the fact that Aristotle and his Stoics held that animals are denied all higher intellectual, aesthetic, and moral feelings. He touched on Descartes and came at last to Darwin, whose work went to prove that animal and human consciousness do not differ in essentials.

Thus in the early days of the twentieth century, said Professor Stumpf, one single, incontrovertible fact to prove conceptual thinking in nonhuman animals would settle the whole heated argument. To anyone who posed the question "Can animals think?" one need only point to Clever Hans—if, that is, one could be absolutely sure of him. If a horse could really solve mathematical problems, as people were saying of Hans, Aristotle and Descartes would be vanquished forever. But could Hans really do all this? The thing had to be approached with caution, and the horse carefully examined to make sure no trickery was involved. What of Mr. von Osten? Those who knew him insisted that he could not possibly be involved in anything shady, such as trickery with a horse. His colleagues approved wholeheartedly of his teaching methods in mathematics, whether applied to human pupils or horses. If anybody could teach a horse to add and subtract, divide and multiply, and extract square roots, said the educational experts, von Osten was the man.

Every day the wonderful horse was exhibited in the paved courtyard that adjoined his stall, in a northern district of Berlin near Mr. von Osten's house. It was not a rural setting; tall apartment houses surrounded it. About noon there would gather in the courtyard a select company of people who had been invited by Mr. von Osten—mathematicians, musicians, psychology professors, and anyone else who had a valid reason for being there. No charge was ever made for the show: it was not, as I have said, that kind of thing at all. The pair of them would enter from the stall, the white-haired Wilhelm von Osten, who was, according to Stumpf, anywhere from sixty-five to seventy years old, holding Hans's halter, and Hans himself, "a good-looking Russian trotting horse" in fine condition, wearing merely a girdle, a snaffle, and light headgear. He seemed very docile, though it was rumored that he sometimes showed flashes of bad temper. Von Osten handled him gently, with soft, encouraging words; he never used a whip. Now and then when a reply was correct he would slip pieces of bread or carrot to Hans. When Mr. von Osten asked questions, if they were of the yes-and-no type they were answered, as has been said, by a nod or a head shake on the part of Hans, who also indicated "up" and "down" with his head. Other questions, which were answered with foot tapping, had to be posed within the limits of the vocabulary with which he was familiar, but it had a fairly wide scope and increased day by day.

It was not only Mr. von Osten who worked with the horse. A certain few others were accepted by him; he was willing to cooperate with them, though he was not always as amenable with strangers. Once he had accepted someone, Hans not only answered questions pertaining to mathematics and musical intervals: he would walk to certain spots in the yard when told to do so, or approach certain persons on command. He would pick up a colored cloth in his teeth from a selection of cloths of different colors and bring it back. But this was only a small part of what he could do. Thanks to Mr. von Osten's code, the horse would transpose concepts into numbers—letters of the alphabet, tones of the musical scale, the names of playing cards—and use them appropriately. He mastered the cardinal numbers from one to one hundred and the ordinals to ten. He could read, though Mr. von Osten taught him only lower-case letters: when placards with several written words were put in front of him and he was told to do so, he indicated one word or another by pointing to it with his nose. He knew the value of all German coins. He carried the entire yearly calendar in his head and could give you the right day for any date you might mention. He recognized persons from photographs. In music he had absolute pitch. There seemed no limit to his powers. But like so many other geniuses, Hans was high strung, nervous, and moody, possessing strong likes and dislikes. Frequently he was stubborn and gave a wrong answer on purpose, though immediately afterward, as if to show he could, he would solve a most difficult problem.

Not surprisingly, Hans captured the public fancy in Germany. He was the subject of popular songs, and you could buy picture postcards of him and bring home little toy Hanses for your children. People wrote to the newspapers about him and controversy raged, some learned men saying that he wasn't really intelligent but merely possessed of a remarkable memory. Much was made of the primitive, illiterate tribes that were reputed to carry everything in their heads. Scholars engaged enjoyably in discussions as to just what constitutes intelligence anyway. Nobody said that Hans was a music hall entertainer: he was considered to be another thing entirely. People who might be supposed to understand horsy matters—such as cavalry officers, the director of the Hanover zoo, and a well-known zoologist—announced publicly that they believed in Hans, that he was sincere.

Nevertheless, there were some who still doubted. Though as far as anyone could see, no signals passed between Mr. von Osten and his horse, there simply had to be signals, they said. Somebody suggested that the mathematics master might be emitting the new x-rays everybody was talking about, which entered Hans's brain. Or was it that mysterious thing called "the power of suggestion"? At length it was agreed that a searching, scientific inquiry should be made into the matter. Mr. von Osten readily promised to cooperate with the examiners, and three men familiar with Hans were named to a kind of committee to assure fair play, along with three observers not so intimate with the horse and its master. Two of them were Professor Stumpf and Oskar Pfungst, who was able to develop a pleasant relationship with Hans. The rules were drawn up. Every now and then one of the six gentlemen was to ask Hans something to which he himself did not have the answers. Also, the interrogators were to be switched now and then.

The stage was set, and the men arranged themselves in the customary formation, with a questioner standing as usual to Hans's right, fortified with bread, carrots, and lumps of sugar. With all due solemnity, a card with a number on it was shown to the horse in such a way that the questioner couldn't see it. Hans tapped out the number "fourteen," though what the card said was "eight." Next time the questioner was allowed to look at the card. It was "eight" again—and Hans tapped out "eight." Unfortunately, when the next card said "four" and the questioner didn't see it, Hans tapped "eight" again. Whenever the man didn't see the card, Hans was wrong; whenever he did, the horse was right. There was no flying in the face of these facts: Hans could not count. But how did he work it?

Reading tests got the same results: whenever the questioner didn't see the words, Hans didn't know them, but he made almost no mistakes when the man knew the answer. So Hans couldn't read, either, but the judges still didn't know how he managed to guess what the man was thinking. He failed the test on calculation of dates; he failed in music.

"Not a trace of musical ability," reported Pfungst severely.

But how did he get the right answers at all? It simply had to be that he picked them up from clues, signals unseen by the watchers. Poor Mr. von Osten became increasingly dismayed as he saw his wonderful horse fail time after time. Perhaps, he suggested, the secret lay in sound waves emitted by the questioner. No, he later decided, that could not be the explanation, because at times it wasn't necessary to speak aloud to Hans: he would start tapping before the question was uttered. Just to make sure, however, they tried him with ear muffs: it made no difference. No, it couldn't be sound waves. No matter how mysteriously he did it, the gentlemen agreed, Hans got hold of visual signals somehow. To try the idea out, they put blinkers on him. Aha! Something different happened immediately. Hans didn't care for this situation, and made what were described as strenuous efforts to regain a view of his questioner, raving and tearing at the lines when they tried to tie him and make him stand still. Mr. von Osten said he had never before behaved like that. Furthermore, his record of answers now became woefully bad.

When they took the blinkers off, Hans calmed down. They asked him questions in the same old way, and now at last they noticed something never observed before—that the horse, when posed with a problem, had a peculiar way of watching his questioner very closely, rather than looking at the subject under consideration, placard or colored cloth. It had never been noticeable when a question dealing with mathematics called for cerebration, as in the extraction of square roots: nobody at such times had noticed at what point the horse's eyes were fixed, but when he was supposed to be reading it was different. Mr. von Osten, called on for help, willingly lent himself to an experiment in which he asked Hans questions from two positions, one at the horse's side and one from behind a canvas wall. The resultant difference was striking. With Mr. von Osten close at hand, Hans got practically all the answers right, but when he was asked similar questions from behind the canvas he was completely at sea. Another thing was noticed: he was at his best at midday when the sun was bright, but as the courtyard darkened his ability seemed to wane.

Now that the committee knew where to look, they soon had their solution. Hans was watching his questioner's inadvertent movements. Having set a problem, the man unconsciously leaned over to count the hoof-taps, and when the proper number had been reached he relaxed. That is, he straightened himself with a slight upward jerk of the head. It was the smallest possible movement, but it was enough for Hans, even when his questioner was Mr. von Osten, whose movements after four years had become so slight and refined that few people could spot them. There was no doubt in anyone's mind, incidentally, of Mr. von Osten's innocence. The mathematics teacher was amazed and grieved, so much so that a rumor gained credence that he had committed suicide out of shame and disappointment. In fact he did not, but he was so downcast up until the time he died in 1908 that it is not too much exaggeration to say he was broken hearted. Professor Stumpf's words may have afforded him some comfort: "Everything would indicate," said the professor, "that we have here not an intention to deceive the public, but a case of pure self-deception." He added that, in spite of the fact that Hans couldn't read or count, he still remained a phenomenon in his quickness to catch on to the people who asked him questions. No other so-called show horse, in his estimation, was Hans's equal, the gestures on which they relied being in comparison "coarse and easily perceived."

Was he all that extraordinary, after all? Probably not, according to a professor of our day, Thomas A. Sebeok, head of the Department of Semiotics and Linguistics at Indiana University. When I commented in his presence that Clever Hans must have been a genius among horses, he demurred.

"No more a genius than any schoolchild," he said. "Can't you remember how you used to watch your teacher's face while you were reciting, to see if you were on the right track? I did, at any rate. That's all Hans's genius amounted to."

Mr. von Osten's champions were not so easily discouraged. Though it might be supposed that the scientific explanation of Hans's amazing talents would put an end to their hopes of communication between our world and that of the nonhuman animals, beliefs in the wonderful are hard to kill. One of Mr. von Osten's friends, a jeweler named Karl Krall, simply refused to accept the committee's judgment. He bought Clever Hans from Mr. von Osten and took him home to Elberfeld, where he set to work creating a school of clever horses—Muhamed, Zaref, and Hanschen. People came from miles around, even from foreign countries, to watch the Elberfeld horses perform, answering questions by tapping their forefeet and all the rest of it. The others, too, seem to have been remarkable animals, but Professor Stumpf maintained that Hans was always the most alert. It took a very sharp eye to discern the signals he caught.

As for those certain other marvelous animals, talking dogs, specimens have been on view ever since the reign of the Emperor Justinian in the sixth century, according to Stumpf. Such an animal's ability "to single out an object on which his master had intently fixed his gaze, was made the basis of a special form of training, called 'eyetraining,' nearly one hundred years ago," he wrote. Usually such a performing dog's master, unlike Mr. von Osten, would not let anybody else work the animal: dog and man, if the act was to be effective, had to stay close together. Yet dogs, too, learn spontaneously to react to slight involuntary expressive movements by their masters. Sir William Huggins, the English astrophysicist, owned an English bulldog named Kepler, who—said Huggins cheerfully—could solve the most difficult mathematical problems: like Clever Hans he could extract square roots and so on. He would bark to answer questions—one bark for one, two for two, and so forth—and was always rewarded with a piece of cake when he gave the right reply. Huggins, however, was not deceived by his wonderful dog. He knew that in spite of his exerting the most rigid form of control over his face, Kepler could read on it when to stop barking.

Professor Stumpf made a telling observation on the subject of dogs and horses. Dogs, he reminded us, quite obviously watch their masters' faces. Who can forget the alert stare with which a dog waits for the next order? Horses, however, are not so easy to read, because their eyes are on the sides of their heads. When Hans watched Mr. von Osten with his right eye, that eye appeared completely dark. He could have been gazing in any direction. Nobody could be blamed for failing to see what he was doing, but the affair was to have wide repercussions in the scientific world. Even today animal research workers are apt to say to one another over the results of some experiment:

"You're sure that wasn't a Clever Hans syndrome?"

The thing may well have discouraged those who wanted to find out more about animal awareness.

CHAPTER 3

Some Statements by Horses

Sometimes the great silence and solitude to which we appear doomed is so oppressive that we tend to welcome too eagerly any hint we think we can detect of an animal trying to get in touch with us by crashing the barrier. We know that those creatures over there do communicate with each other in various ways, dogs by scent and posture, horses by puffing, kicking, and nudging. There are many other behaviors that we can observe. But what we look for is something more, something perhaps on a higher level than the commonplace, on which the animal merely begs for food or safety or amusement. The search for this kind of communication leads people down many different streets into—in some cases—some very odd places indeed. That is why it comes as a delightful surprise to be told by a solid researcher like Karl von Frisch that the curtain has been pulled back a little. Even the most skeptical were at last forced to admit that von Frisch's study, which he published half a century after the affair of Clever Hans, proved the possibility of understanding communication of a different species. It is odd that this breakthrough was accomplished by observing, not horses or dogs or cats, the most familiar of domesticated species, but honeybees.

Von Frisch decoded the honeybee's "waggle dance." What everyone had always assumed to be the purposeless wandering of the bee was, it now appeared, a remarkable method of communication from one bee to another about food sources. It was complicated and yet—when it had been explained—beautifully simple. A worker bee, having discovered a food source, returns to the nest to tell the others about it. This is done by dancing and at the same time "buzzing" by shaking her wings rapidly. The dance consists of running about in a distinct pattern, as if she were inscribing a figure 8, at the same time waggling her body. The middle line of the 8 gives the direction the other bees must fly to aim straight for the food, and also tells how far away it is: the length of the middle-line run indicates this distance. One of the startling elements of this dance is that the bee, if it is outside the nest and on a horizontal surface, simply aims in the right direction, but if it is inside and on a vertical surface, the straight-line run follows exactly the right angle to the perpendicular. In one race of bees, the Carniolan, a run lasting a second means that the food source is five hundred meters distant, and a two-second run means that a flight of two thousand meters will be necessary. The directions are always, as Harvard professor E. O. Wilson says, respectably accurate.

Is this speech that can be compared to human language? At first, Professor Wilson grants, it seems very much like it, but he has reservations. "The great dividing line in the evolution of communication lies between man and all of the remaining ten million or so species of organisms," he wrote in his *Sociobiology*. "The most instructive way to view the less advanced systems is to compare them with human language. With our own unique verbal system as a standard of reference we can define the limits of animal communication in terms of the properties it rarely—or never—displays." The waggle dance of the bee has symbolism in its straight run ("the ritualized straight run"), and bees using it can generate new messages with it. Just the same, he thinks the waggle dance is severely limited when one compares it with human language.

This is true, but most of us don't compare it with human language. The facts seem staggering enough by themselves.

Bird song, insect signals, animal noises, and behavior generally have all been and are still being investigated. But science works slowly and painstakingly; it is only natural that a number of people should try to take short cuts, feeling that they have already plumbed the mysteries. Such nonscientists write to the papers about their dogs or cats or horses, animals that show striking talent for understanding what their masters are saying. Some of them, according to these accounts, can even speak for themselves, in their own way. At least one man, an Englishman named Henry Blake, has gone further than this and written a whole book on the subject *Talking with Horses*. Blake's father, a farmer, made his children ride bareback until they were seven, and Henry was so much at home with the animals that as a schoolboy he successfully handled a dangerous mare that had killed her former owner. She proved a useful animal, and never again attacked anybody. In time Blake and his father gave up farming and merely trained difficult horses.

Blake does not believe in breaking horses in the customary manner, feeling that the barrier to riding an animal for the first time is in the mind of the rider, not that of the horse. Time after time he has himself ridden new horses without any kind of trouble. He might have gone on indefinitely in his slightly unorthodox but not unheard-of fashion if he had not suddenly felt that there is more to getting along with horses. It happened one day at a market, when he seemed somehow to get a message: somebody was saying, "For God's sake get me out of this." Blake knew immediately that the source of the appeal was close by, "a sixteen-hand dirty brown thoroughbred horse, who was as thin as a rake." He just knew it, that's all. He bought the animal, Weeping Roger, on the spot.

From the very beginning, Blake says, they seemed to have an affinity for each other. It was Weeping Roger who taught him how much power a man can have over a horse if he really applies his mind to controlling and handling it. With another animal—since Weeping Roger was, of course, already broken—he tried a new method, working intensively at gentling him. On the second day he rode the horse without incident, and after that he rode it for an hour every morning and another hour every afternoon. In a week, he assures us, he and his father took the new animal to play polo, and in a slow chukker did very well. Ten days after training began, the horse played in a fast chukker and never put a foot wrong, though, as Blake points out, it normally takes at least two years' schooling to make a polo pony. He was tremendously excited. What was this mental control he had stumbled on?

He knew that there are people who seem to have a special bond with certain horses. His three-year-old daughter had a little pony named Darwi that was too small for Blake to train, so he enlisted the help of a neighbor's daughter named Doreen. For three or four months Doreen handled Darwi until the pony was nicely settled. In fact, the relationship was so successful that whenever Darwi got lonely or bored at night he would get out of his field and make his way to Doreen's house, two and a half miles away, to bang on her window with his nose until she came and talked to him. The beautiful friendship was interrupted when, in the course of time, the Blake child outgrew the pony and Darwi was passed on to another small child, whose people moved out of the neighborhood, taking him with them. Not long afterward the Blakes also moved, to Wales.

Seven or eight years later, at a Welsh gymkhana (or riding competition), they ran into Doreen again, when she hailed them joyfully, in a loud voice. Blake was chatting with her, neither of them paying any attention to a burst of whinnying at the far end of the field, when suddenly they saw a pony running toward them, the child on its back trying vainly to control it. The pony was Darwi, of course. He ran around the track and came to a skidding halt by Blake and Doreen, to greet the girl with every sign of excited happiness: he had known her voice after eight years.

Such stories as this led Henry Blake to think even harder about communication between horses and people. He tended to believe that horses above all other animals are in tune with us humans. As a boy he had often wondered why it was not as satisfactory to ride other species than horses. He tried riding cattle and sheep and pigs, but it wasn't the same: the other animals simply didn't cooperate in the way horses do. Just why, he asked himself, does a horse cooperate? At any rate it does, he reflected, and has done so for ages of time. He thought of the examples in prehistoric art of men riding horses, and the representations of the same exercise in Greek sculpture. Xenophon once said that horses are taught not by harshness but by gentleness; Blake approved of this dictum. He admired the American Indians, whose method of talking gently to their horses, slowly coaxing them to accept blankets on their backs as a prelude to actually riding them, is similar to his own.

In his circle, of people interested in horses to the point of obsession, there is much talk of an almost legendary Irishman known as Sullivan the Horse-Whisperer, famous for his success with recalcitrant horses. He would take the most savage animal, a notorious killer, into a barn with him, shut the door for an hour or so, and come out all in one piece, leading the horse, which had miraculously become quiet and tractable. He never told anyone, even his own sons, how he did it, but Blake thought Sullivan, too, probably used methods like his own. Talking gently, without stopping, he would put his hand on the horse's back and imitate the movements of a mare nuzzling her foal. Then, as the animal relaxed, he put both hands on its back, soothing it and talking to it until a bond of sympathy had been created.

Some trainers blow into a horse's nostrils, occasionally with powder in which is hidden a so-called taming oil, which allegedly quiets the animal immediately. According to an Englishwoman, Barbara Woodhouse, who caused a furor by telling in a television interview that she blows into a horse's nostrils and thus "talks" to him, a similar practice is followed in South America. Henry Blake believes that the method does work, at least sometimes, because horses when they meet blow through their nostrils in greeting one another. Not, however, that this greeting is always amiable; it all depends on degree. Hostile horses blow hard, but a gentle blowing is friendly, like that of a mare caressing her foal, when she puffs so gently that one can hardly hear her. If you try this method, Blake advises, keep your hand on the horse throughout because horses like physical contact and depend on it when they are frightened. However, he is convinced that there is something more than this to getting on well with a horse. He firmly believes in extrasensory perception, and to prove it he cites a happening at a show in which he was riding a horse with which he was very much in tune. During rehearsal he had tried out one or two new turns, but there was one he avoided, thinking it too complicated to risk without more time for preparation. Yet in the middle of the act the horse went on and did exactly the move Blake had half planned, finishing with an action his master had never before tried with him, galloping toward the president's box and stopping dead at the last minute, rearing up on his hind legs.

One would like to hear Professor Stumpf's comments on all this. Certainly had Blake been riding Clever Hans, even that perspicacious animal would have found it impossible to read his rider's signals. But he could have *felt* them, perhaps—changes in knee pressure, or the handling of the reins, even breathing.

ESP or no, Blake thinks that not many trainers can hope to achieve so fine a degree of understanding. However, there are capable horsemen who don't go in for his way of training but who have developed nevertheless a good deal of communication with their mounts with ordinary spoken commands such as "Whoa" or "Giddap," heel touch, pulling or releasing the reins, caressing in praise or striking in anger. And the horse makes signs to us, greeting human friends with a whicker or slumping to show he is tired. But these, Blake maintains, are merely the rudimentary gestures: he is capable of many more. Blake talked these matters over with his wife, who is equally dedicated to horsemanship, and resolved to compile, with her help, a dictionary of horse language, or rather messages, a message being defined as any intention, threat, inquiry, feeling, or statement made by a horse. One thing that made this project very difficult, he admitted, was that such messages are not cut and dried, but are subject to variations. In the three hundred horses studied by the Blakes during their compilation work, they noted thirty different ways in which the animals said welcome, and as many variations on the question "Where is my breakfast?" On the average a horse's message can be put in six to ten ways. "Come and drink," is one of these, and another is "Come and fight." Some other ordinary, or common, statements are "Don't do that," "Don't go away," "Don't leave me behind," "Help!" "I am boss," "I am frightened," "I hate you," "I love you," "I suppose I will have to." Like other foreign phrase books, Blake's gives the distinct impression that horses, like tourists, are insecure.

But his is not an ordinary phrase book or dictionary, because many of the sentences are perforce accompanied by descriptions of behavior. For example, the message "Who are you?" as used by horses on meeting for the first time, includes sniffing or, more usually, blowing (as Barbara Woodhouse has testified), as well as the simultaneous carriage of head and tail. One feels that a supplementary moving-picture tape would be useful.

Henry Blake would no doubt enjoy meeting a contemporary, Mrs. Elisabeth Mann Borgese, journalist, animal trainer, polyglot, and incidentally a daughter of Thomas Mann, because her mind, like his, is open. There is nothing of the skeptical scientist about Mrs. Borgese, as any reader of the book *The Language Barrier: Beasts and Man* (1965) will realize. She studied the records on Clever Hans and the accounts of Karl Krall's collection of wonderful horses at Elberfeld. She noted that Krall had arranged that his horses, unlike Hans under the management of the mathematics teacher, were really placed in seclusion, walled off from questioners, so that nobody could argue that they were getting visual signals. Nevertheless, many observers insisted, they tapped out correct answers to questions in the old way. Mrs. Borgese wondered if Hans had not perhaps been unfairly accused. Certainly a lot of people who went to interview the Elberfeld horses came away convinced that their talents were genuine, and Mrs. Borgese did not dismiss the possibility that these people were right. Hans and the others may have known what they were doing.

There was also an airedale named Rolf, she found, who had learned to talk to people by the Elberfeld method, tapping his paw in numerical code. Nor was Rolf unique, because more horses and dogs were constantly being initiated into the mysteries of the technique. Between the two world wars, Mrs. Borgese calculated, there were at least one hundred and five animals communicating in this way—sixteen horses, eighty-eight dogs, and one cat. Somehow that cat gives one pause in one's first impulse to laugh the whole thing to scorn. It is hard to imagine even one of the species bothering to extract square roots.

Mrs. Borgese grew more and more intrigued by the subject. In the 1930's, she read, a Milanese trainer named Mar had induced his dog Bonnie to use letter cards with which she spelled words, pulling them with her nose from a pack and arranging them on the floor. Another Italian, Signor Tagano, wrote a book about Bonnie entitled *Your Dog Can Write*, and published it in 1939. In it he said that Bonnie's vocabulary contained more than five hundred words. During the troubled years that followed soon after this effort, Bonnie was forgotten, but after the war public interest in literate animals was revived, and Mrs. Borgese found herself in direct touch with the subject when she was sent by her employers to report on a wonder poodle named Peg, who lived in Brescia and was reputed to be adept in mathematics and spelling. She used letter cards as Bonnie had done, but on occasion would bark to express herself, rather than patting the ground with her foot. To test her, Mrs. Borgese devised a method that would obviate the possibility of the dog's watching her reactions in Clever Hans fashion. She had brought with her a picture book. Now, holding the book's face away from her, she opened it at random so that Peg, and not herself, could see an illustration. When it had been exposed to the dog's gaze, she marked the place, closed the book, and put it into her briefcase.

"Did you see what I showed you in the book?" she asked Peg,

who replied with three barks, meaning yes. Peg then sorted out the proper letters and spelled, on the floor, the word *cavalli*, which means "horses." Opened at the marker, the book revealed a picture of two Belgian cart horses. As Mrs. Borgese points out, the dog had not only got the right animal but knew the difference between singular and plural. There was no doubt in her mind now that Peg's barking was no mere music hall trick. Interspecies communication was a fact, and Mrs. Borgese longed to prove it to a doubting world.

CHAPTER 4

A Bad Dog

A lot of people demand proof when so surprising a thesis is proposed, but Mrs. Borgese was not unique in her obvious desire to believe, with a clear conscience, that certain fairy tales are true. There was a significant reaction to a book that came out in 1961— *Man and Dolphin*, by Dr. John C. Lilly—and that starts out with the sentence, "Eventually it may be possible for humans to speak with other species." So many people bought and read *Man and Dolphin*, so many letters were written to the press arguing for or against, that one cannot doubt that among the public is a large percentage of souls akin to Mrs. Borgese and Dr. Lilly, who yearn to believe in that possibility.

"It is true!" cried one side, and "It is only a trick!" cried the other. Those in between refrained from coming down hard for either truth or falsehood, murmuring, "Maybe so, maybe so, but in any case those animals must be very intelligent to learn such complicated tricks." Which, of course, brings us into the old argument: What is intelligence? What is a trick? I suppose one good definition would be that a creature is intelligent if it does something on its own, with a purpose in view, whereas when it does a trick, its act is stereotyped and merely calls for a reward.

Dr. Lilly said that if we grant the possibility that man can someday speak with other species, it is evident that the other species should have an intellectual development comparable to our own: that is, it should have a complex brain, nearly as large in comparison with the body bulk as ours is. (The human brain, measured according to that standard, is the largest in nature.) The elephant, too, has a large brain, actually as well as comparably, but there are manifest difficulties in working with it. Elephants, being large and strong, are not easy to handle, let alone easily available to scientists in America. The same fact applies to the larger whales, which might otherwise fill the bill. But dolphins, which are almost little whales, are another matter, especially the bottlenose dolphin, *Tursiops truncatus*, which lives offshore for the most part in shallow, warm waters. Dr. Lilly, whose interests were primarily in the field of neurophysiology, began taking an interest in these animals when he saw some of the exhibitions that attract the public to Marineland, Florida, and similar aquatic parks—Marine Studios, Seaquarium, and the like. Dr. Lilly noted that the average brain weight of *Tursiops* is 1,700 grams, not so very far, as these things go, from the human brain weight of 1,450 grams (though the percentage of brain weight to body weight is lower, of course), and that dolphins are of far more manageable size than are their relatives the whales, or elephants. Evidently, too, people can get through to dolphins to an amazing degree, since the animals can be taught such human activities as basketball and dancing—at least, dancing on top of the water—and can even give vent to noises composed of clicks or a kind of "raspberry" sound. This is hardly strange, since the whale, relative of the dolphin, can make noises that are very complicated. Recordings of the song of the humpbacked whale astonish and delight the human listener. There can be no doubt that dolphins communicate with each other by means of the sounds described by Lilly and others.

Further: "Each animal puts its blowhole out of the water and emits the sounds in air, rather than under water as he is accustomed to do," said the doctor. "They accommodate to us rather than we to them."

At first he approached his investigation as a neurophysiologist rather than a psychologist: after all, that was the discipline he had been trained in. He mapped the brains of his dolphins and learned what areas gave what results when electrostimulated. He worked on the concept of physiological reward and punishment for various actions, turning in one direction or another, plunging, swimming after and catching a floating ball, and so forth. As he put it: "We can push a button and cause a brief controlled period of a rewarding pleasure. . . . Conversely, we can push another button, which controls another place in the brain, and cause intense punishment (fear, anger, pain, nausea, vomiting, unconsciousness, etc.)." Such a method is all worked by wires, and the results, whether the subjects be dolphins or monkeys, are comparably effective. Even so, said Dr. Lilly, setting up understanding between dolphin and man is not simple: their way of life is vastly different from ours, not only because they live in water but because their aims and activities are so unlike. In search of food or a more agreeable temperature, dolphins swim thousands of miles in a few days, and seem none the worse for it. They don't have to store food as we do; they don't need clothes or shelter; they don't have to resist gravity. Because of freedom from gravity they do not sleep as we do: in fact, they ought never to become quite unconscious, because they don't breathe automatically as we do, and breathing stops altogether in the unconscious state. A dolphin deeply asleep would not surface every so often to take in oxygen, and that means he would drown. Lilly once saw a case in point. A dolphin being lowered into a tank struck his head against the side of the pool. He passed out, and sank immediately to the bottom of the pool, whereupon the other dolphins raised him to the surface and held him there until he regained consciousness and began again to breathe. Dolphins do take care of each other in this way, which is probably why there are so many cases, well authenticated, of people being saved by them from drowning. In Wilson's language in his book *Sociobiology* the animals are "altruistic."

With seven other investigators, Lilly went to Marineland and arranged to borrow five dolphins from the authorities, then the party set to work. They found out a good deal about dolphin neurophysiology, but in the process all the animals died, and within two weeks, at that. The Marineland people were upset, and told the scientists that this was not the way to treat intelligent, friendly, playful animals. Thinking things over, Dr. Lilly was inclined to agree. But he really had found out a lot about dolphins, he explained; the five had not been sacrificed in vain.

It was when he started to work on another dolphin that everything suddenly happened. To begin with, he was amazed that the creature seemed to understand what he was doing when he applied stimulation to the brain's pleasure areas. For some reason Lilly had thought of putting a switch within reach of the dolphin's beak, while the animal watched the procedure closely. Then, with what he describes as amazing rapidity, the dolphin got the idea and pressed the switch for himself. It was this same dolphin that Lilly found the way to make vocal. Whenever the stimulation stopped, he emitted a great burst of different noises, producing them simultaneously in a jumble—"Whistles, buzzings, raspings, barks and the Bronx cheer-like noises." Lilly said he had the eerie feeling that the dolphin had understood the whole program from the beginning. Something even more eerie happened shortly afterwards.

He and his assistants had been laughing in the laboratory about some trifle, when all of a sudden from the dolphin's blowhole (from which many of the noises emanated) came a burst of sound similar to that laughter. Could it be that the dolphin was trying to imitate them? Surely not . . . but for some time they had been taping its noises anyway, trying to induce repetition of whistling. They had found that the animal did indeed respond to the recording of its own noise by replying with more whistling, but evidently something even more interesting was now happening. Playing over the tapes, they found that the dolphin, "in a very tense and quacking sort of way," had copied, or mimicked, sections of what Lilly said to his assistants during their work. One of the phrases so mimicked was "T R R [train repetition rate]," said by Dr. Lilly very slowly and distinctly to the secretary in dictation, "is now ten per second."

At this point on the tape the dolphin could be heard to say, "T R R," in its own peculiar fashion; to use the doctor's words it was "a very high-pitched, Donald Duck, quacking-like way." The team continued their experiments, now watching carefully for more imitations, and found that the dolphins could copy a human vocalization and compress it at the same time, so that the imitation lasted for only a second or two, and was pitched very high. The humans could make it out only by slowing down the tape. For this imitation, dolphins evidently employed the rasping or creaking noises commonly used when they searched for food by sonar, and these came from the blowhole, but another mechanism for producing sound was in the larynx and airway situated below the blowhole. Thus dolphins could make two series of noises at the same time. Unfortunately, after a long session in the laboratory, the animal known as Number Six, from which the experimenters had first learned so much, sud-

denly had an epileptic seizure and died. Dr. Lilly was sorry, but not
sorry enough to stop the work. He acquired more dolphins.

Number Eight added to the list of his species' accomplishments.
Whistling in reply to Lilly brought him a reward, so for a long time
he whistled without other stimulation to get more food, going higher
and higher until he was out of human hearing range, and the only
way people could tell that he was emitting noises was from the sight
of his blowhole, still twitching. It seems incredible, but according to
Lilly Number Eight soon caught on to the difficulty, and thereafter
kept his whistling down, within the right acoustic limits.

On the whole, however, said Dr. Lilly, in spite of the accommo-
dating nature of the animals, one could not ignore the obstacles that
stand in the way of complete communication between species so
unlike in their habitat. "In the air we live in we can hardly hear
what these animals have to say," he observed, "and in the water
they can hardly hear what we have to say." Therefore, to pursue his
investigations more freely, he finally set up a laboratory out of doors
in the Virgin Islands, acquired some more dolphins from a professional catcher in Florida, and opened for work, in 1960. Before the
work in the Islands a few years later, he and his assistants learned a
great deal more about the capacity of dolphins to imitate and
produce what he calls "humanoid sounds." Sometimes the tapes
took a lot of adjusting before anything could be made out, and one
had to accustom oneself to the strange noises even then, so that not
everybody not working in the laboratory could honestly say that
they could hear very much that was significant. Dr. Lilly, however,
was not discouraged. He was confident that he was on to something
of world-shaking importance in interspecies communication.

They had a dolphin they called Elvar whose short, quacking
noises began to sound very much like human words. Elvar was
particularly good at imitating the sounds made by one of Lilly's
assistants, a woman named Alice, who sometimes addressed the
dolphin in affectionate baby talk, rather like that used for infant
animals—"soothing sounds and attention-getting sounds, clucking
with one's tongue," as Dr. Lilly wrote. Elvar copied these noises.
They were extremely primitive copies, the doctor admitted, obviously made by a vocal apparatus quite different from ours, but they
were copies nonetheless. Elvar and other dolphins often imitated
human laughter, whistles, Bronx cheers, and even certain simple

human words. The outstanding example he cited was that of a dolphin named Lizzie. It was late one day, and Lizzie was not well. They had been force-feeding her with a stomach tube. Dr. Lilly was reluctant to leave her, and one of his people, probably the cook, called down from the house to say that it was late and if he didn't hurry he might spoil his dinner.

"It's six o'clock!" the remonstrator said at last, very loudly. A little later, equally loudly, Lizzie said—well, what did she say? Opinions differed. On tape it sounds like a poor copy of the words "It's six o'clock!" but to Lilly it sounded more like "This is a trick!" They were never able to tape a repetition of the phrase, because in the morning Lizzie was dead.

In the late 50s Dr. Lilly moved the laboratory to Florida and carried on just outside Miami, but after two years there he closed it down altogether. Financially it seemed impossible to continue; besides, he came to the conclusion that he had no right to go on hurting dolphins: he had grown too fond of them to maintain his detached scientist's attitude. He taught for a time at Esalin and thought about other matters than dolphins, but there is a rumor afoot that he might start his researches again. I have not been able to authenticate it, but if it is true it won't be long before we are hearing of Dr. Lilly and his dolphins again.

The Lilly experiments and his book about them have had an explosive effect on the public. One might have thought that they had been waiting all their lives to hear about dolphins talking: everywhere people were clamoring to hear more about the subject, or writing philosophical treatises inspired by Dr. Lilly's book. Marinelands and Sea Worlds, which had always been looked on as rather childish entertainment, suddenly found themselves catering more to adults than to children. Rather more sinister was the report that some officials in control of American defense had begun to train dolphins—or, at least, to try to train them—to serve as an arm of the fighting navy by carrying mines. The idea was that they should act as kamikaze divers, blowing up enemy ships that might in due course threaten our shores. Indignation ran high when the plan became known: the people of America had just learned to love dolphins, to look upon them as the world's hope for bridging the gap between us and all the rest of the animal world. There had even been talk of communicating with visitors from outer space through the benevolent interception of our friends the dolphins. It was indecent to think of perverting these lovable animals for such a revolting use, said people, and they may well have been right.

Another effect of Dr. Lilly's researches was on Elisabeth Borgese, who remembered the gifted dogs she had interviewed. If dolphins could communicate, why couldn't these dogs? Peg, the genius, had died in 1963 at the good old age of fifteen, but there must be other brilliant canines, if only one could find them. Mrs. Borgese tried earnestly, traveling through various European countries in her search until she got a new idea. If other people could teach dogs to read, she reasoned, she herself could do it. To be sure, Peg had been a poodle, and poodles are reputed to be quicker-witted and generally more clever than all other breeds: a lot of them can be seen as performers on the stage. But there is no immutable law that all clever dogs are poodles. As it happened, the Borgese family had pet dogs at home in Italy. They were English setters and very intelligent, all four of them. Mrs. Borgese went home and got to work.

Healthy, bright, inbred, and spoiled is how she described the setters in her book: one of them, Jinxy, could open every door in the house. Because she resented being left alone with the other dogs when the humans went out, she would take revenge by pulling books from the shelves and throwing them to the other dogs to chew up—a fairly widespread problem for dog owners all over the world, as a matter of fact, but to Mrs. Borgese, obviously a fond mistress, it denoted special intelligence. But it was Trixie that Mrs. Borgese found most interesting. She needed love and often demanded it, coming to stand in front of her mistress to be combed and groomed. At the end of the ceremony Mrs. Borgese was supposed to say, as a sort of ritual, "Trixie, now you are beautiful. What a beautiful dog!" upon which Trixie would bark and dance joyfully. When Mrs. Borgese didn't have time to administer the whole treatment, the "ritual words" were enough to send the dog capering. Sometimes, too, she would come into the house from play limping on three legs and holding up her forepaw as if it had been injured. There, too, ritual was expected. "Poor Trixie, poor Trixie," Mrs. Borgese was expected to say, patting the dog and blowing softly on the paw. "There now, it's okay," she would finish, and Trixie always galloped away on four paws, her injury forgotten.

One more sign of Trixie's unusual intelligence, as noted by her mistress, was founded on a common illustration of dog behavior: if Mrs. Borgese announced that they were going for a walk, Trixie barked joyfully and ran around in a circle, chasing her own tail in a way familiar to dog owners. Soon, however, she elaborated on the mannerism: whenever she had to go out by herself, especially at night, she would chase herself around in a circle until some human got the message and opened the door for her. Such ready, intelligent adaptation of a symbol seemed a good omen for the new project. Mrs. Borgese was confident that some one of her dogs would prove capable of emulating Peg.

After carefully reading the words of other students of animal behavior, she set to work. The beginning of the training incorporated a method, already used by trainers, that depends on an animal's capacity to recognize certain visual symbols and distinguish between them. One of the chief experts in this field is Dr. B. Rensch, who has worked with various animals, including elephants. Mrs. Borgese put each of her four dogs to a test by locking herself into a room with him or her. As a first step, a plastic cup containing a bit of food was placed on the floor, and the dog was permitted to scoop the tidbit from the container and eat it. The procedure was repeated several times, until Mrs. Borgese was sure that the sight of the container would imply to the dog the idea of food. The next time she put down the loaded container, she covered it with a plastic plate. At first the dog, having been well brought up, did not know what to do, and whined helplessly, but when he got the idea that it was all right to knock off the plate—in fact, that he was supposed to do it— he went at the task enthusiastically. When this simple experiment had been completed a score of times with each dog, everyone was permitted to rest.

During the following days, the lessons became, little by little, more complex. First, the plate covering the cup was decorated with a large black mark, until the dog had had time to get used to it. Next he was presented with no single covered cup, but two, one empty and the other baited. The plate over the empty cup was marked as usual with a black mark, but the one hiding the piece of food bore two black marks instead of one. It took the dogs a long time and many tests before they learned the significant difference between the signs, but once they had learned they never forgot. Mrs. Borgese stresses that it wasn't the food they worked for but the joy of the game, they were well-fed dogs.

Now at last, she felt, it was time to move on to speech. She taught her pupils that when she said, "One," they were to select the one-mark saucer, and when she counted, "One, two," they were to choose the two-mark plate. By the end of four weeks she was confident that all the dogs could count to two, but they balked whenever she tried to go beyond it with "One, two, three." So she gave it up for a while and turned to discrimination tests. These, again designed by various animal psychologists, aimed at teaching the difference between negative and positive, by means of contrasting symbols. Copying the methods of Rensch, the German professor, she adopted, among others, the symbol of a black cross against a white background for "yes," and a black dot, which stood for "no." In one week the Borgese dogs had learned to recognize nine such pairs of symbols, eighteen patterns. Fortified by her success in this, she tried again to persuade the dogs to count to three. Two of her pupils fell by the wayside on this one, but the other two passed the test, though one of them was not very good at it. The less adept student, too, was discarded, leaving Mrs. Borgese with one young dog, Arli, to work on. Alone, Arli forged ahead. When Mrs. Borgese painted whole words on his plates he learned to recognize them: soon he could "read," after his fashion, the words DOG, CAT, ARLI, BIRD, BALL, and BOWL—though, she admitted, he had no conception of their meaning. In "arithmetic," as Mrs. Borgese termed the game she made up, Arli could choose the larger of any two numbers up to five, but he was never very happy in the higher reaches of counting.

"It is very difficult and tiresome for a dog to learn a sequence of five numbers," Arli's teacher wrote apologetically.

Still, he made steady progress. He learned to pick out in proper sequence the letters for CAT and DOG when they had been jumbled together—though, as she again reminds us, he did not know what he was spelling. Now, she decided, he was ready to learn typing. The Olivetti Company, interested in the work, donated to the cause an electric typewriter, and she had it adapted. It would have been impossible for any dog larger than a Chihuahua to hit one key at a time, so she got an imaginative electrician of her acquaintance to build extensions to the machine: these provided that most of the keys, though not all, were represented on the superstructure by much larger circles, just the size of the plates Arli had worked with in his primary school. To begin with they were blank, but Mrs.

Borgese painted the *A* key with a large black *A,* led Arli to the typewriter, and told him to hit it with his nose. He obeyed. The resultant click startled him and he recoiled, but his teacher tempted him back to the machine by holding a bit of raw meat under the *A* key, and soon he was used to the noise and ignored it. Using the same stratagem, after painting correct letters on the right keys, she persuaded her pupil to touch *R, L,* and *I,* until after a few trial runs Arli was typing his own name. After six months he could type twenty words, though admittedly he made a lot of errors, especially when writing a word he knew well: when he reached such a word he had a way of typing one letter over and over, as if to show that here, at least, he could do it right. Two years after he was first introduced to the machine Mrs. Borgese confronted him with a typewriter similar to his save that there was no marking on the keys. Arli went straight ahead and typed the words he knew, evidently, she said, because he had the letters and sequences as thoroughly "in his nose" as we humans, using the touch system, have them in our fingers.

Sometimes Mrs. Borgese forgot her own honest dictum that Arli couldn't really read or write, in the sense that he didn't know the meaning of what he was typing. She found it necessary now and then to remind herself of this, because otherwise she would be getting false notions. As a scientific explorer, she reflected, she must refrain from romantic interpretations of the dog's feats. Nevertheless, he did something one day that shook her badly.

Arli had just undergone a long, trying plane journey, and he wasn't in a good mood when she started the daily lesson. He lay stretched out on the floor and ignored her preparations until she called him sharply to come along and take dictation. Unwillingly he got up and approached the typewriter, and she began, as so often before, to dictate, letter by letter, the sentence GOOD DOG GET BONE. As always, Arli dropped his nose to the keys, but his mistress could see that he wasn't hitting the right letters: the first one he typed was *A,* and he seemed to have some idea in mind, so she let him have his own way just to see what would happen. When he had finished the sentence, she read it incredulously:

A BAD A BAD DOOG

My God, she thought; have we broken through the barrier? Did Arli know what he had just written? Then calmer thoughts intervened, and she decided the phrase had been the result of chance and mechanical training. BAD DOG was one of the combinations Arli had in his nose: once he had started it would naturally finish itself ... but what had started it? She went on wondering, hopefully. Afterwards she sometimes urged the dog to go ahead and write without dictation. He did it, but reluctantly, and they never made much headway. It was clear that he preferred to take dictation and hated to think up words by himself. When she left him to his own resources very long, he would protest by whining, hitting the keys with his paws instead of his nose, and generally slowing down.

Arli's case again brings up the question of trick versus intelligence. Some people would certainly maintain that what he was doing was a trick, but we then find ourselves in that gray area toward which such definitions always drift. Sea lions that balance balls on their noses, bears that dance, dogs that walk on their hind legs are no doubt doing tricks: these are stereotyped actions instilled in them through training that is attuned to physiological possibility. Seals naturally balance very well, bears are habitually at home on their hind legs, and dogs likewise, as Dr. Johnson observed, can be trained to go against their nature and rear up in bipedal position for a time. But when we go beyond these simple cases, doubts occur. Is it a trick when a child learns to multiply? One might say no, arguing that the process of multiplication demands reasoning power. Yet many children, if not most, don't know why multiplication works, though they can do it. One may with truth say that they have learned a trick, because the logic of mathematics is something the student grasps later. In fact he may never grasp it, though as an adult he can add and subtract well enough to keep a checkbook balanced.

We tend to call a dog or a chimp or an elephant "intelligent" because it learns quickly to do what humans teach it. Perhaps we should say instead that the animal is very adaptable. And yet there are some animals in which one can see, in their dealings with us, a primitive readiness to reason. The gray area is large, because it includes, as it should, young human children learning algebra as well as puppies being taught to take part in circus performances. Where, in this mist, are we to draw the line?

Moreover, the word "intelligence" is full of pitfalls, as witness Professor Eric H. Lenneberg, linguist and author of *The Biological Foundations of Language*, who, when asked which is the more intelligent, a dog or a cat, replied, "Comparing the intelligence of different species is comparable to making *relative* measurements in *different* universes and comparing the results in *absolute* terms. When we say that a cat is more intelligent than a mouse or a dog more intelligent than a cat, we do not mean that the one can catch the other by superior cunning but that one solves *human* tasks with greater ability than the other."

CHAPTER 5

Silent Elephant

Arli, thought Mrs. Borgese, had gone about as far as he could go, and in her mind she cast about for another animal that might take her beyond what he had done. She had read a good deal of the works of Professor Rensch, the animal behaviorist particularly involved with communication: in fact, it was to him that she owed much of the technique used in training her dogs. But Rensch's specialty was elephants, and it is not as easy to study these animals as to observe dogs. Nevertheless, Mrs. Borgese was not daunted. She went to India and paid a visit to an elephant-training station near Madras, where a number of the animals are taught to haul, carry, and stack lumber. Much impressed with elephant sagacity, she borrowed a young animal for a time and worked with it, putting it through the same exercises she had employed when teaching Arli. The infant proved very quick at learning the symbols. Unfortunately Mrs. Borgese lacked the right sort of special typewriter, and couldn't continue her research beyond the primary cup-and-saucer exercises, but she came away from India convinced that much in this line could be done with elephants, if only one had the right tools for teaching.

She fully realized that there was nothing novel in this idea, and so should we. Centuries before Professor Rensch proved that elephants are quick witted and can figure things out for themselves to an astonishing degree, trainers were claiming that their animals had extraordinarily retentive memories and great powers of understanding. In 1510 King Manoel of Portugal read a book by an Italian about travels in India, in which the sagacity of elephants was described so excitingly that Manoel made up his mind to acquire a number of the beasts. (Portuguese ships often visited India, where elephants could be bought.) The king built an elephant stable in Lisbon and lost no time filling it, eventually falling so much in love with his animals that he made his architect put an elephant into the decorative whirligigs of a tower he was erecting in Belem, where it can be seen today. Manoel often exhibited his elephants in shows or processions through the city streets, where they would pause every so often to do tricks for the amusement of the spectators. When Pope Leo was enthroned in 1513 and the kings of Christendom, according to custom, sent missions to Rome with gifts for the new pontiff, Manoel outdid all the others by including among his offerings an elephant named Hanno, accompanied by his Hindu mahout.

However, the business of sending the elephant did not go at all smoothly. Hanno balked at getting aboard the ship detailed for his passage at Lisbon, and people said it was because of the Hindu mahout, who had fallen in love with a Portuguese girl and did not wish to leave Lisbon. He had told the elephant terrible things about Rome, the gossips said, and that was why Hanno refused to embark. King Manoel when he heard about the incident was very angry: he threatened the mahout with severe punishment unless he reassured his charge and got him aboard the ship without further incident. Hanno did get into the ship ultimately, and arrived safely in Rome. There he had a smashing success with his new master, Pope Leo, who spent so much time with the animal that criticism was evoked. Martin Luther expressed the public discontent in 1520 when he published a scathing article about popes who waste their time playing with elephants.

Though some skeptics have been known to argue that the elephant doesn't really comprehend language, Mrs. Borgese, for one, has no doubt on the subject. To the argument that the elephant obeys merely in reaction to gestures and punishment—the mahout pushing his animal with kicks or stabs from the ankus—she retorts that though some of the elephants she observed were indeed trained in this manner, others were able to work without riders, simply obeying spoken commands. For some time she watched the animals clearing a mountainside in the jungle where trees, already felled and stripped, were scattered over the ground for moving. One of the logs was prepared, with a hole drilled through one end of it and a chain put through the hole. A rope was attached to this chain and then an elephant ridden by a mahout approached it. The mahout directed his elephant to pick up the rope, drag the log uphill, lift it over an obstruction, and so on until they reached the top of the slope, where the elephant dropped the rope. A man waiting there untied it and removed the chain from the log, whereupon the elephant draped rope and chain over his tusks and retraced his steps down the hill to work on another log. In the meantime a second elephant picked up the first log with his trunk, made sure it was evenly balanced, and carried it across the crest of the hill to the other side. This animal was not being ridden, but obeyed instructions from a man standing on the ground nearby. The elephant set down the log in what was obviously the correct position, then kicked it downhill to where two more elephants were waiting. They picked it up and stacked it neatly on a pile of other logs, always in obedience to human directions.

When the day's work was finished, all the elephants were taken to the river for their daily bath. There the mahouts stood on the riverbanks or in the water shallows and told their charges when to roll over, when to sit up, and when to squirt water over themselves, while men in the shallows vigorously scrubbed the giant animals. It was a combination of vocal and gestured orders that seemed to work very well, evidence of a truly astonishing degree of communication between man and animal. It is nearly all one-way traffic, with man talking to elephant and elephant hearing and reacting, but not having much to say for himself. To be sure, that is not the elephant's fault. His species doesn't make much noise, for one thing: they squeal, they trumpet, and sometimes they make a muttering sort of sound that Mrs. Borgese calls "humming," but there is little variety in their spoken vocabulary. Of course, there are other ways to communicate: Mrs. Borgese heard of, without actually meeting, an elephant in Kerala who was trained to write on a blackboard, with a piece of chalk held in his trunk, the quite legible word WELCOME. It is not easy to decide whether this achievement proves anything more than that the elephant who did it had an excellent memory for muscular effort in the way to move the chalk. More to the point, if we can believe the story, is that elephants do associate sounds with signs. Mrs. Borgese tells us that when an elephant of her acquaintance was shown a chain bearing the printed label "Chain," he remembered the connection and would pick up not only the actual chain when told to do so, but, if it was not there, the label instead. Some elephants have been taught in this way to recognize several words for familiar objects, and will pick them out on request without making mistakes. Is this reading? At any rate, it is as close to reading as the dog Arli ever got.

I wondered if I should not perhaps get a look at a few elephants for myself, and judge the possibility of communicating with them. Because I was in the United States, a long way from either India or Africa, the first step seemed obvious: to visit either the circus or the zoo, and it wasn't circus season. Fortunately, I happen to know a zoo where there are a lot of elephants, in Portland, Oregon. Nobody knows just why the Portland Zoological Garden should hold the American record for breeding Asian elephants, *Elephas maximus;* Oregon's rather damp, cool climate is not in the least reminiscent of India or Ceylon, but so it is, an unsolved mystery. My friend Dr. Phil Ogilvie, at that time director of the Portland zoo, was not complaining, at any rate.

"We're proud of our record," he assured me as we entered the gates and walked along the wet pathway. Both of us wore raincoats: there was a thunderstorm going on at the moment. "For one thing, if we go on having an oil crisis in this country, elephants may turn out to be very useful. They can pull or push most heavy weights, and they're intelligent about it. Who knows? If we're successful at teaching them all those lumber and timber tricks they learn in Thailand, we can save an incalculable amount in energy."

As the rain stopped, we approached a huge barn, outside the doors of which an elephant was standing, attended by a man and a girl. "This is Packy," said Phil after greeting the humans. "He was born here a pretty long time ago—twelve years, I think. The zoo got his mother, Belle, from Thailand, and she's still here, getting old, of course. I have a list of all the elephants we've bred here if you'd like to see it. We've had fifteen births so far, and the infants are all over the place, in Los Angeles, Brookfield, in Chicago, Vancouver, Oakland—oh, a lot of zoos. Most of our animals, the ones we imported, came from Thailand, but we have one from Cambodia. They're all the same subspecies, as you probably know, but we've begun branching out just lately. We've got an African elephant and hope to find a mate soon, so with luck we'll have one breeding pair of those. Nobody's bred an African elephant yet in America. It should lead to some fascinating comparisons even if they don't have offspring. The African looks quite surprisingly big, but then Thai elephants do run a little smaller than the true Indians. So do the ones in Ceylon."

The girl attendant got up on the elephant. As the dust from the hay bales sheltered in the barn made me sneeze, we strolled to another part of the building complex, through numerous enormous, echoing empty stalls separated from each other by ponderous doors. Phil explained that the doors are electrically operated and very strong. It seemed a fine place for elephants. We came at last to a kind of amphitheater, large, airy, and oval in shape. It had no audience room, however, and no seats anywhere. Along the edge of the floor rose strong steel bars at intervals through which human beings, but not elephants, could easily slip. At one corner of the room, between two of the bars, stood an edifice the size of a telephone booth. A group of young assistants was clustered around it, so I went to investigate. The booth, I observed, stood on little wheels like casters, and contained a narrow room in which one of the assistants was standing, fiddling with some sort of control. You could see him through the little door. On the other side of the little shack, the side that faced the big arena, an opaque glass screen was inserted. It looked something like a switched-off television set. Under it were inset two round glass discs like large buttons, one on the left, one on the right, and beneath these a small cup was set into the frame.

The man in the booth said he was ready, and a signal was given to people who were hovering near one of the stable doors. Then, slowly and with dignity, an elephant was led in, a female, who needed no coaxing from the man who led her: she looked eagerly at the screen, as if she knew what it meant.

"She's one of our old girls," said Phil to me. "Now watch. This is a light discrimination test."

The man inside busied himself, and the disc on the right suddenly glowed. The elephant lifted her trunk and pushed at it, and a sugar cube rattled into the cup from inside the booth. She picked it up, popped it into her mouth, and looked again at the machine. The process was repeated, and she got another lump of sugar. On the third test it was the disc to the left that glowed: the elephant duly pressed that one and got her sugar, but the fourth time, as we all waited expectantly, neither disc was illuminated. The elephant tried anyway, pushing at the right one.

"That was wrong," said Phil. In vain did the elephant's trunk finger feel into the cup: it was empty.

She had a few more trials before she was led away. On the whole she did very well.

"They first rigged this thing up about eight years ago," Phil explained. "It was for a psychological study by some people in the local university, in combination with the zoo. They ran variations on the theme, of course: sometimes they reversed the experiment, for example, so that the animal was supposed to select the dark disc rather than the light one, but it was all very much the same thing fundamentally. Then interest in the results waned, after I don't know how many tests—it was before my time. Or maybe they ran out of funds, or wrote their dissertations and went away, I don't know, but it was all forgotten until lately, when we found the apparatus again, pushed into the back of a storeroom, and Hal Markowitz and I decided to try it out. It was all beat up, but we got it back in shape and brought out the elephants. We wanted to see if it's true that they have such good memories: three of the females had taken the tests in the old days, so we tried them out again, and do you know? They did remember. It was eight years ago, but they remembered. Our first elephant took only six minutes to go through the whole routine, and she made only two errors. That's quite something, isn't it? The two others didn't do so well, though, and we couldn't understand why until it occurred to us that they might be having eye trouble. We made tests, and that was it, all right. Belle is almost totally blind and the other isn't much better. We wondered, naturally, why we hadn't ever noticed it before. The answer is that the others help them: they compensate so beautifully that until we made the tests we never saw anything wrong. Thanks to their companions they move around, eat, and do almost everything exactly as if they could see."

The first performing elephant had been led away by this time, full of sugar and looking rather smug, to be replaced by a younger, smaller animal named Temba, who arrived surrounded by a number of young people in blue denim. These were described by Phil as trainees and students, who gave a hand in taking care of many of

the animals. Temba, he said, was about four years old and hadn't
been born in Portland. She came direct from Thailand and had been
in America about a year. Directed by a number of urgent voices, she
now took up her position before the booth and waited. A light
flashed: she tapped the button on the right, but that was the wrong
one, and when she fished in the cup she found nothing. The next
time she got it right, and so it went on for a time, but clearly the
game was too slow for Temba. Behind that opaque window, she
must have told herself, was an unlimited store of sugar. Why waste
time tapping silly buttons? She lifted her trunk into the air and felt
behind the top of the booth.

"Temba! Temba!" shouted all the blue-denim brigade, and one or
two of them tugged hard on the chain attached to her collar, but
Temba was strong, and ignored them. She stood up a little on her
hind legs and again investigated the booth's top with her trunk.

"Temba!" they shouted. This time, with concerted effort, they
managed to pull her down, and the game was over. The disappointed elephant allowed herself to be led away, and went out with
shuffling step.

All these experiments had been carried out, as far as the elephants were concerned, in silence. The only sound I heard one of the
animals utter was while I was watching a young one being exercised
by a keeper in a field near the stables. Shackled to his hindquarters
was a large log, which he was supposed to tow along in approved
working-elephant fashion. The log didn't impede his steps, but he
seemed bored by the whole affair, perhaps because it was so simple.
What he wanted was to wallow in mud, and there was plenty of that
if one looked for it. Now and then, as he made his rounds, he and his
leader had to splash through a shallow puddle, and each time they
reached such a place the elephant let his legs relax, trying to get his
body down to the glorious mud. Each time he made the attempt the
keeper tugged him upright, and the elephant squealed.

"They never make much noise, do they?" I said to Phil, and he
agreed. That baby elephant was about the noisiest of the lot, he
added, and that was because of his youth. Elephants squeal as a sign
of protest, trumpet in anger, and mutter because—one doesn't know
just why they mutter. For the most part they communicate with
each other, and with us if we know how to understand, by gestures
made with ears, eyes, trunks, and bodily attitudes. "It's the way
they put their feet," as one zoo director told me. In conclave they

move about so purposefully that one can understand the belief, picked up and used by Kipling, that they dance. Deep in the Congo forest I have seen much-trodden swampy places, out of the sun, which it was not hard to believe were what the Pygmies assured me they were, elephants' ballrooms, but even the Pygmies did not claim that the animals made much noise in their merrymaking. Considering their size, elephants are remarkably silent creatures. They must consider us positively verbose—but then, as I have said, they have other means of communication. Evidently it's not that we cannot communicate with them, but that they can't talk to us. As in so many other relationships, the man-elephant connection is for the most part a one-way affair. Any elephant keeper knows that his animal understands him, and understands a good deal better than most humans can understand the elephant.

CHAPTER 6

Complex Behavior

"Language is extremely complex behavior," wrote Dr. Eric Lenneberg of the Harvard Medical School, in his book *Biological Foundations of Language*, "the acquisition of which, we might have thought, requires considerable attention and endeavor." He was discussing in context the fact that healthy deaf children two years or older get along very well in spite of their deficiencies in speech. "Why do hearing children bother to learn the system," he continued, "if it is possible for a child to get along without it? Possibly because the acquisition of language is not, in fact, hard labor—it comes naturally—and also because the child does not strive toward a state of perfect verbal intercourse, normally attained only two years after the first beginnings."

In support of his thesis, he cited cases of children who for one reason or another were temporarily unable to make sounds, but who, when the hindrances were removed, produced sounds typical of their ages, instead of going back to make up the stages they had lost. Also, many children up to the age of two years, admitted to hospital because of severe neglect by parents or guardians, were at first very unresponsive and seemed to be retarded in speech as well as in other functions; yet after a few weeks of care they began to make all the sounds typical of their age. Older children do not come out of this sort of induced torpor so readily, but even so, many of them do at last begin to talk fluently, in accordance with their age level. Though they have not practiced speech and language like normal children, one cannot say they have not been learning.

"They simply do not choose to respond," Lenneberg concluded. He did not say so, but since the temporarily afflicted child can jump, as it were, from a state of no babbling to a more advanced stage of speech, one might hazard the guess that infant babbling is not really necessary as a prelude to speaking. This idea runs counter to the theory expressed by Cathy Hayes in her book *The Ape in Our House*, which describes how she and her husband raised Viki, a female infant chimpanzee, and tried earnestly to teach her to talk. The experiment lasted from 1947 to 1954 and was never really successful, though on the way the Hayes couple learned a lot of interesting things about chimpanzees. For example, in her early days Viki never babbled, a fact Mrs. Hayes thought significant of their failure, for she accepted the theory that babbling is an important part of speech learning in a human child. According to Lenneberg, of course, this is not necessarily true, but he observes that there is a definite period in the child's life when it picks up languages easily—a statement that few of us would dispute. However, he holds that this phase is limited in time. When it comes to an end, usually at the onset of puberty, the child's facility in language learning vanishes, and afterwards he or she must undergo the same trials we adults suffer when we set out to learn a foreign language. It's not impossible to do, but the knowledge does not come so easily as once it did.

At this point, one might begin to think of the wolf children of Midnapore, a fact of which Dr. Lenneberg is fully aware, for he himself mentions them, but with reserve. The Midnapore children's story has long exerted a fascination over the public, like the similar case of the so-called wild boy of Aveyron, about whom a motion picture was recently made. Besides, there is Kipling's Mowgli. So perhaps we had better look again at the Indian "wolf girls," always remembering, however, that not everybody believes they existed.

In 1920 a Christian missionary of the Bengali town of Midnapore, the Reverend J. A. L. Singh, went out on one of his routine soul-saving expeditions in the jungle, pursuing his vocation among the aboriginal tribes that lived in scattered villages in the wilds. He and his wife also ran and supported an orphanage at the home

mission. The expeditions he described in his diary were rather
elaborate. He and various companions would take with them as
many as thirty volunteers from the villages, pagan people not averse
to listening to him preach Christianity as long as they could hunt
between sermons. Sometimes the party stayed all night in the
jungle, their bullock carts and draft animals with them in a huge
circle of burning logs: you could see the wild animals in the firelight,
wrote Mr. Singh in a diary that was later published (*Wolf Children
and Feral Man,* by Singh and Zingg). He and his companions
carried guns; the villagers, being poor, depended on bows and
arrows and spears, as well as a few large drums with which to scare
the animals. It was always a happy time, said Mr. Singh, and he
traveled like this year after year.

In October 1920, his party stopped at the village of Godamuri,
where they heard a strange story about a man-ghost that had been
seen in the jungle seven miles away. During the past five years more
than one person had encountered the horrible apparition, said the
villagers. It had a man's body but a dreadful, huge head. They were
terrified of it, and begged Mr. Singh and his companions to shoot it.
Mr. Singh and the others visited the spot, and he instructed the
villagers to construct a machan, or hunter's blind, in a nearby tree,
from which they would have a good view of the ghost. Then the
travelers went and borrowed a pair of field glasses, and lodged for
the night in a cowshed belonging to one Chuna of Godamuri. In the
morning they went over and inspected the terrain, finding a large
termite mound that stood as high as a two-story house, "in the
shape of a Hindu temple." Holes led into the mound, though there
were no longer any termites there.

That night at 5:00 P.M., after dark, the ghost hunters climbed up
into the machan and waited. An hour later they saw a full-grown
wolf trot out of one of the holes. (Mr. Singh called the animal a
wolf, though a footnote explains that it was not a true wolf of the
European species, but a jackallike creature characteristic of much
of India.) This animal was shortly followed by another of the same
size, and after it trotted a smaller one. Then two cubs appeared, to
be followed by the ghost. It was, Mr. Singh admitted, a hideous-
looking thing: its body was that of a human being, but the head
seemed to be "a big ball of something covering the shoulders and
upper portion of the bust." A face was visible under this ball—
which later proved to be made of matted hair—and it was clearly

human. After the first figure, which was not, to be sure, the size of a human adult, came another smaller figure, just the same in appearance. Mr. Singh observed their eyes, which were very bright, and decided that the creatures were human beings. Before emerging altogether, the larger ghost paused, elbows on the ground around the hole, and looked around, then leaped out and followed the cubs. The smaller creature did the same in its turn. Both ran on all fours. When walking, they placed the palms of their hands flat on the earth. Occasionally they would crawl on hands and knees, but when they wanted to run they stood on hands and feet and ran very fast, so that it was very hard to catch them.

Members of the party aimed at the ghosts and would have shot them if Mr. Singh had not insisted that they must not. The creatures were human, he said, and must be captured, not killed. Having examined them again through the glasses, the others admitted that he was right, and they all went back to the cowshed to make plans for the capture. When it was proposed to the villager Chuna that he dig into the termite mound, he flatly refused. He was very much afraid. Singh's party did not insist, but left Godamuri and moved on to another village where nothing had been said of ghosts; there they hired men to dig out the mound, and brought them back without notifying the first villagers. The digging was done on the morning of October 17. As soon as it began, two wolves ran out and fled into the jungle. The third wolf, a female, did not run away, but stayed on the spot to fight it out. Mr. Singh hated to kill her: it seemed to him, he said, divine that she had kept the two strange creatures (though no doubt she had originally brought them back to the den as food for her cubs) and brought them up. However, his helpers quickly dispatched her with bows and arrows, and that was that.

The diggers opened up the mound and found the strange creatures and the cubs huddled together in what Mr. Singh called, vividly, a "monkey-ball." A sheet was thrown over the lot of them and each of the animals swaddled in a separate covering, then separated. The cubs were given to the villagers to sell, and the human children—for they were merely children—were carried back to Godamuri. Mr. Singh put them in an enclosure in Chuna's courtyard for the time being, with a pitcher of water and a plate of rice: Mr. Singh himself had to move on, to save souls elsewhere.

Unfortunately, Chuna and his family were so afraid that soon after Mr. Singh disappeared they all ran away, leaving the children to starve. When the missionary came back after several days they were nearly dead. He had to give them cloth soaked in hot tea to suck, as they wouldn't eat; later they took milk, and he was able to put them into his bullock cart and take them back to Midnapore. After the children had been washed and their hair was cut off, they looked much better. Judging by their teeth, the Singhs decided that the elder one, whom they named Kamala, was about eight years old, and the younger, Amala, a year and a half.

They were mute at first. The only sound either made was a peculiar cry, described by Mr. Singh as beginning with a hoarse voice and ending in a thrilling shrill wailing, very loud and continuous: no doubt we would call it howling. Almost every night they howled about three times: at ten o'clock, at one in the morning, and again at three. Kamala's voice was stronger, Amala's thinner: "But both had a fine thrill of reverberating notes, very high, and piercing. It could be heard from a good distance. . . . At first it startled everyone in the Orphanage, but we soon became familiar with it."

For a little time the new children showed more than a passing interest in only one of the other orphans, and that, no doubt, was because he was too young to walk, but only crawled like them. However, they soon frightened him by getting angry and scratching him, and he avoided them thereafter. They didn't even try to get friendly with the others, but crouched together all day in a corner of the communal room and did nothing; musing, as Mr. Singh said. They were never left completely alone, but though the other children played and ran about and chattered, the wolf children did not attempt to make friends or join in.

"They were looking for the wolves and the cubs," commented Mr. Singh. If any human approached them they shrank back, sometimes showing their teeth. Three months after the capture they frightened one of the children who had been guarding them by biting her, and slipped past her, and got outside the compound, hiding under some lantana bushes and making no sound while the Singhs looked for them. At last they were found and brought back to resume their morose ways. This went on for weeks and months, with no improvement. Mr. Singh took notes from time to time of their habits: how they could smell meat at a great distance, like animals; how Kamala tracked down some raw meat where it was being prepared and tried to get hold of it, growling as she did so, and how they could sit or squat on the ground but could not stand up straight. At first they ate from plates on the floor like dogs, never using their hands; they lapped milk or water from this position as well.

Soon they began to look for Mrs. Singh whenever they were hungry, going in to the room where she was and hovering about until they got food and drink. Amala learned to make a sound when she wanted milk, saying "Bhoo, bhoo," but Kamala only touched her lips with her tongue at such moments. At night they slept huddled together like puppies—that is, when they slept—but they were restless and often got up to prowl in the dark. They tore off whatever clothes the Singhs put on them, giving in only to loincloths that were fastened around them. They hated to be washed.

All this being so, it is surprising to read that the expression on their faces was bright and pleasant. Indeed, Kamala was always smiling except when she was angry or afraid, when suddenly she appeared very ferocious. Neither child ever laughed outright. But sometimes when all the orphans were out of doors they would play together, "running about, the two children jumping on feet and hands, looking at one another in a different manner altogether."

The first hopeful sign that they might be socialized came in March 1921, when a cow got loose in the garden. The other children ran in all directions, but the wolf children went straight to Mrs. Singh for safety. When they were ill only she could feed them; they closed their eyes and ignored all other people who tried to help. She was attempting constantly to accustom them to the sound of speech, talking to them all the time, though they seemed only to be bothered by her conversation. She also took babies and played with them close by, prattling and playing little games. Though the wolf children pretended to pay no attention, they sometimes looked sidewise at the infants. One day Mrs. Singh distributed cookies to the babies and also offered them to Kamala and Amala. They pretended not to notice, but she put the cookies down near them and went away, and they took them. Soon they were accepting such things direct from her. She also massaged them with oil for an hour every morning.

In September of that year, 1921, the children fell ill with dysentery complicated by parasites. They were seriously ill, and Amala died on September 15. Kamala tried vainly to wake the little girl up, and then, for the first time, she showed signs of emotion: two tears trickled down her cheeks. For five days after the younger child's body was buried she still sat in a corner, her face to the wall. Then, little by little, she showed an interest in the outer world, leaving her corner to crawl over to some young goats that were living in the house. Now and then she took a couple of the little kids in her lap, and after a while she actually tried to talk to the goats, forming words like a young baby just learning to prattle. She petted them affectionately, and, said Mr. Singh, "at times her face brightened up so as to approach a human smile for a passing second."

When the kids outgrew the house, Kamala turned her attention to other living things—chickens, who did not return her affection, and a hyena cub that was there for a while. But the cub died, and when there was no other animal for her to become friendly with she had to depend, in the end, on her fellow humans. Mrs. Singh stepped up the massages, administering them morning and evening. This seemed to have had a good effect on the child both physically and mentally. One day as Kamala lay on her bed she made room for Mrs. Singh to sit down near her, took the woman's hand, and put it on her chest as if asking to be rubbed. This was toward the end of 1922. It was a few days later that Kamala began watching Mrs. Singh play with the babies. When one of their toys happened to roll near her she would pick it up—especially if it was red—and run away with it in her mouth, as if frisking.

The Singhs began encouraging her to stand up, or at least to kneel up. Food was served to her not on the floor but above her head on a stool, so that she had to kneel up to get it. They gave her a pillow to help support her waist. In a few months she was walking a little way on her knees and no longer needed the pillow. Next, the Singhs arranged benches in a square around a higher table on which they put a plate of tempting tidbits. Kamala watched the babies, who propped themselves on the benches to reach for the plate: it was too high and they failed, but Kamala waited until she thought no one was watching, then she, too, tried, and succeeded after learning to straighten her legs for a moment. Next day the Singhs took her out into the garden with the cat, which often played with Kamala. The cat climbed a tree and Kamala attempted to do likewise. She didn't get very far, but she did straighten her legs to sit astride a branch.

After another year there was no danger that Kamala would run away: she became afraid of the dark like the other children, and at night would stay close to the Singhs out of doors. Now it was taken for granted that she come out with the babies when they went for their outings. By this time she could stand up, though she never learned to run bipedally. When she was in a hurry she dropped to the old position of all fours.

In 1924, when she suffered another attack of dysentery and had to stay in bed for a few days, the wolf girl amazed everybody by suddenly beginning to pronounce words. In all, forty-five words were added to her vocabulary during this time. To be sure they were poorly pronounced, but she clearly knew what they represented. After that, she kept trying out new words, talking to herself as she played. Then she began singing the words, humming so loud that at times she could not hear when she was addressed. That year Mrs. Singh went away to Calcutta for a few days, and Kamala moped until she came back. When Mrs. Singh came in the girl ran joyfully to her on hands and feet, turning around and around and rubbing against her, then walking proudly back to the house with her adopted mother, jabbering eagerly the whole way. She often depended on signs when words failed her: once, when another child hurt herself in the garden, Kamala ran to get Mrs. Singh and managed to get the idea across without saying anything.

During the next few years, Kamala's progress was steady if slow. Not only did her language ability impr.ove, but she was definitely one of the group of children at the orphanage, sharing their duties and games. Her eating habits, too, became more human. Once, when they found a dead chicken and everyone expected her to pounce on it and eat it, she didn't, and when asked if she wanted it she said, "No, no." When dogs barked at her she seemed afraid, and went a long way around to avoid them—a great change, commented Mr. Singh, on both sides. She developed and learned. But in September of 1929 she fell ill of an ailment none of the local doctors could diagnose, which made it impossible for the Singhs to accept an invitation from the Psychological Society of New York to bring the wolf girl to America and go on a lecture tour with her. Mr. Singh had long since given up trying to keep her concealed, and many papers had published reports about her: interest was high. Kamala died on November 14, at the age—if Mr. Singh's estimation was correct—of between sixteen and seventeen.

This, the most detailed account in literature of wild, or feral, children, has been studied, edited, and commented on by Dr. Robert M. Zingg, associate professor of anthropology at the University of

Denver, in *Wolf Children and Feral Man*. Dr. Zingg also included in his book a number of other names of what he describes as extreme isolation: Kaspar Hauser; the wild boy of Aveyron; the bear children of Lithuania; and many others. Some of the accounts are reasonably credible, but others are very doubtful. A few of these children may well have been idiots from the start of their lives, and cast out on that account. Others, of course, were not necessarily retarded before they experienced enforced isolation from their kind. In any case, Dr. Lenneberg dimisses the subject abruptly:

"Descriptions of children supposedly reared by wolves or growing up in forests by themselves are plentiful, but none is trustworthy. The only safe conclusion to be drawn from the multitude of reports is that life in dark closets, wolves' dens, forests, or sadistic parents' backyards is not conclusive to good health and normal development."

Without doubt, this is only one man's opinion, but if the wolf children of Midnapore were genuine, as many believe, there are tantalizing questions to be answered. For one, how did they communicate with their wolfish foster families? It is impossible to answer, though Amala, at least, was so young that she could not be expected to develop any kind of code or language, either with wolf or human. Kamala's was a different case altogether.

Charlie, a chimpanzee at the Washington Park Zoo, Portland, Oregon, prepares to make the signs for ''look/see'' as he regards himself in his mirror.

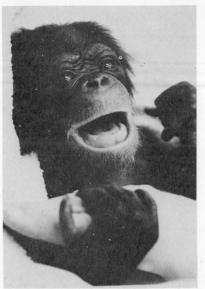

Delilah, a month-old infant in the nursery of the Washington Park Zoo, smiles at one of her human handlers. Chimps that are raised by humans seem to begin socializing with people sooner than those raised by their natural chimp mothers.

Washington Park Zoo

Washington Park Zoo

Charlie does not need human sign language to show that he is alarmed and enraged. Here his hair is standing on end, and he reinforces the message by barking.

Lucille Ogilvie, wife of former Washington Park Zoo director Dr. Philip W. Ogilvie, in a sign-training session with Delilah and Bathsheba. Mrs. Ogilvie began by giving the "eat" sign, Bathsheba imitated her, and now the chimp is getting something to eat as a reward. Their signing hands are still in the "eat" configuration.

Above: Leah, aged about two years, asks Charlie, about six, for food. Both her hands appear to be in the "eat" configuration, apparently to ensure that Charlie gets the point. *Below:* He does and she is fed.

Susie "greets" Dr. Philip W. Ogilvie, former director of the Washington Park Zoo, Portland.

Dr. Ogilvie talks gently to Susie about the paint pots and paper. He has made a small mark on the paper to show her what can be done with paint and brushes.

Susie goes to work and creates a painting herself.

Charlie, as a baby, makes the sign for "drink" and *(below),* in response to the candy in the trainer's hand, the sign for "sweet."

Bathsheba signs "berry" for a raisin.

Bathsheba signs "orange."

An older Charlie signs ''apple.''

Charlie's handler signs, ''You want chase?'' or ''Chase?'' to which Charlie, body poised to take off running, replies, ''Chase!''

Charlie signing "listen," for his wind-up musical toy radio, as nursery school children watch.

Charlie, also known to his friends as Charles the Mighty and Magnificent, signs "up." He wants to be picked up.

CHAPTER 7

"He Knows When You Talk Nice"

In a collection of papers written by various experts in the field of what Professor Thomas A. Sebeok of Indiana University has called "zoosemiotics," entitled *Animal Communication*, each writer has tried valiantly to define what he means by the term, and if no two of them actually agreed on a definition, at least they provided a most stimulating lot of theories. As Professor Sebeok himself tells us, "Speech is the principal, but by no means the only, mechanism whereby communities are knit into social organization via a systematic flow of messages exchanged over interpersonal communication channels." There are other mechanisms, he points out: such "sense modalities" as learning, vision, touch, smell, taste, even temperature. Gulls communicate with each other by means of postures and movements as well as calls. The male crocodile roars to attract the female, but also attracts her with the fountain of water sent up by the roars, and a musky smell from the glands in his mouth.

Should all this, I asked myself as I read, come under the heading of communication? I decided that it could if communication be defined as the sending and receiving of messages, getting through from being to being, animal to animal. The signals may be so subtle that one doesn't immediately realize what is happening, or one may be so accustomed to them that they are scarcely noticed. I thought of an example: two babies are carried past each other in the street.

Without their making a sound or an obvious gesture, it becomes clear to the observer that they recognize each other as members of the same club, sister ships, because each infant brightens and turns its head to watch the other baby in passing.

I wanted to understand the viewpoint on these matters of people who have to do with animals in an intimate way—trainers, perhaps, or zoo keepers. As luck would have it, I didn't have to look far for an example of the first class. Ringling Brothers Circus was in town at Madison Square Garden, featuring their famous animal man, Gunther Gebel-Williams, and I quickly got permission to interview him.

Excited roars filled the air, filtering in through the doors that lead backstage at the Garden, the same air redolent of hay and animal smells generally. Outside the office men ran back and forth carrying long pieces of iron. They shouted at each other, not in excitement but merely to make themselves heard. On the other side of the enormous room, between pens full of animals, a tent had been erected and was evidently serving as home for a woman and a small towheaded boy who sat on the floor as if he were in the country, playing idly with a piece of cloth, while his mother hung out a little washing. This, it seemed, was the temporary home of Gunther Gebel-Williams, and this was Gebel-Williams himself entering the office, a slight, very good-looking, golden sort of youth with piercing blue eyes.

"I understand that you always talk to your animals," I began.

"Yes, I do," said Gebel-Williams. "I think when talking you come a little closer. An animal can tell when you talk how you are going to treat him. He knows when you talk nice, he knows everything's nice. You know when you are in your house, with all your things around you, you feel safe. Well, the animal feels like that when he hears your voice, I mean my voice. I have this feeling the animal knows how nice I am when he hears me: it's not the words but the sound of it. And I know what he is saying to me, that is, if I know him very well for a long time. When you make a special sound like 'shhh' to a tiger and he does it back to you, he is saying it is a nice day. It is friendly; it says 'I know you,' things like that. Only the tiger among the animals makes that kind of greeting, but it is the same talking with the elephant. If I say, 'Oh, you are a nice girl, what a nice girl you are,' right away she is friendly. And the horse, I think he also knows the difference between hard and soft talking."

As if it had just occurred to him, he added, "Tigers make that kind of noise to each other, too, if one meets another. That is how they can tell they are friends. We will go out later, into the big room, and you can try it on some of my tigers and see if they answer. When they don't answer it means they aren't interested just then, that is all. Of course, with the cats it isn't always that you can be nice with them. Sometimes with leopards or panthers, especially when they are new to the training, they get angry and make yowling noises and try to get at each other. When that happens I can't just let it go by, so I say, 'Here, that's enough now, stop it.' I might stop the training altogether until they calm down. Often at the beginning of training lions and tigers go 'Aaargh' at each other. They are showing off. Don't work with them that day, and next day will go better."

"I've heard that trainers find African and Indian elephants different to work with," I said. "Do you think it takes longer to train an African?"

"Not really longer, but it's a different approach," he said promptly. "The African is shy. The Indian elephant says, 'Okay, I enjoyed that, stay here and maybe in five hours I'll do it right.' The African says, 'Let me do it my way, I'll do it anyhow, but don't hold me too close.' It took me a long time to find that out: I've been working with elephants seventeen years and it's only lately I know. Sometimes the African wants to do something by himself; all right, I let him. They are different like—like a cat and dog. But I have now an African who works very nicely in the ring with an Indian."

"Have you ever worked with bears?"

"Not really. I like an animal that when you look at his face you know what he is thinking. The bear, he doesn't look you in the face and you can't tell. The ears are covered—you can see nothing. He goes—" Suddenly Gebel-Williams averted his face and made a sort of growling roar, pawing the air with his arms at the same time. "No good," he said emphatically. "Maybe he is nice, maybe not, you don't know. The Russians have the best bears I ever saw, but the training, I think, must be very rough. Maybe the Russians can do it, I don't know—nobody can see what they do. They couldn't do it here. Anybody who wants to can see us training. I have seen chimps and monkeys riding ponies in Germany, but I think first they were very young—a chimp can get very strong when it grows big—and also I am sure they were doped. I wouldn't like to work with apes. All the time they have their minds on how to get the upper hand. No, I wouldn't like it. Cats are best."

Then we went into the big room and made the rounds of the animals, resting in their cages between performances. All the way around, Gebel-Williams talked to them, soothingly, lovingly, unceasingly, and they obviously liked it very much. A tiger sneezed at me pleasantly.

"He is a fantastic animal," said Gebel-Williams. "Always you have the feeling he likes you. Look at him!" The tiger regarded us benevolently from his royally reclining position. "Wuzza, wuzza, wuzza," said Gebel-Williams as he patted the animal, until the band, suddenly striking up, drowned out his voice.

In further search of those who understand animals at close quarters, I visited the Arizona-Sonora Desert Museum outside of Tucson, and asked many questions of Charles L. Hanson, curator there of birds and animals. I realized, I told him, that one couldn't always expect hard-and-fast evidence, but wasn't there something of the sort to be spotted at times?

"There are lots of things we don't understand about animal signals," he admitted. We were sitting in his office, a very pleasant place surrounded by low-lying shrubs and, here and there, a tree. A pair of small ferrets were making themselves free of the place, scampering about, wrestling with each other and sometimes taking mock refuge in the bookcases. They emitted shrill, tiny squeaks, like rusty machinery, as they boxed and scampered. On the opposite side of Chuck Hanson's desk, with a desk of her own, sat the zoo's education supervisor, a young woman named Mrs. Doris Ready. Every so often she had to stop listening to the conversation to answer a telephoned inquiry about school visits or publications.

"Why animals react as they sometimes do to strangers is a puzzle," said Hanson. "I know there are criticisms of observations made in a zoo, on the ground that the situation is so artificial that such observations have no validity. I don't think that's true. Even in an artificial situation they would still reflect intelligence levels, communication levels, and behavioral patterns that are characteristic and valid, and they should be given serious consideration because, after all, they simply cannot be made in the wild. It's the only opportunity we have to document. I think most people are beginning more to accept such knowledge. When we see the relation between bobcat and human here, for instance, there's no reason to suppose it differs radically from communication between bobcat and bobcat in the wild. Certainly we shouldn't just ignore it, though the interpretation, of course, is always open to question. I must admit that here at the museum we are emotionally involved with our animals, which precludes objectivity, but even so the observations themselves are valid. Interpretations by other people are quite all right, as long as one doesn't throw out the material."

We talked for a while about relationships between animals in captivity, animal to animal as well as animal to man. Mrs. Ready has raised at home a kit fox, one of the little animals that live in the western desert. (The Arizona-Sonora Desert Museum gives no house room to exotic animals: all its inhabitants are native to the region.) The fox has lived in the house ever since she was very small, and has made one or two strange friendships in the process, the most familiar one being the Ready cat.

"An ordinary house cat?" I asked.

"Yes, just a cat. When the fox was just a few weeks old, and the cat was about a year and a half, they met for the first time. I wasn't at all sure of the cat's reaction, so I held on to it while the fox investigated, and then I held the fox so the cat could relax, and after a while it seemed safe to let them both go. For a while, I forget how long exactly, nothing happened, but one day when they were both out in the yard all of a sudden the fox began doing her thing behind a bush, dodging and posturing, ears laid back, making a kind of giggling sound. Then suddenly the cat jumped into the air and started chasing the fox, playing like a kitten, chasing the fox, and that was just what the fox wanted. They played around like that all over the place. Sometimes they chased each other in the house, fox after cat. It was the fox that finally began the grooming thing, starting to nibble, nibble, nibble on the cat, and the cat didn't care for it at all. He kept backing up."

Chuck Hanson explained to me: "Kit foxes' teeth are admirably constructed for grooming. The muzzle is quite big, and at the tip there is just room for the two front teeth they use in the exercise. They love to groom each other."

Mrs. Ready took up her story. "Well, the fox kept doing it, and then one day, I don't remember just when it was, the cat started grooming the fox back again. Nowadays when the cat doesn't feel like it he just walks off; other times they groom each other for as much as ten or fifteen minutes. Sometimes you have a wet fox and a wet cat. They take turns: after the fox has groomed for a while she seems to feel it's her turn, and she stands there sort of leaning toward the cat with her eyes half closed. If he isn't interested he just walks off."

"The only time the cat objects," Chuck said, "is when the fox starts grooming his ears, but she can't be stopped: it's an important part of the kit fox's social behavior. I told Michael Fox about it—you know who he is, don't you? The man in St. Louis who studies canine and cat behavior. Well, I told him about this very interesting intergeneric social behavior, and he was intrigued by it. It didn't really surprise him, though. He said if you deprived a kit fox of all other animal association, it's so extroverted and so social that it would even groom an elephant."

"We had another weird combination of friendships," said Doris Ready ruminatively. "We had a vulture—"

"Where did you get that?"

"Oh, we raised it from an egg," said Mrs. Ready. "When it was about half grown I tried to introduce it to the cat, and the cat split. It wasn't seen again for the rest of the day. But the fox took to the vulture right away, and they would chase each other back and forth, back and forth. The fox always initiated the play, and I'm not sure if the vulture really enjoyed it. It's hard to tell: a vulture doesn't have quite the expression on its face that a fox has."

"A kit fox will relate to absolutely anything," declared Hanson. "That's one reason it's an endangered animal. It thinks the whole world is its friend."

"Except for one animal," said Doris Ready. "I once brought home a skunk and put it on the floor, and the fox walked right by as if it wasn't there. All the animals ignored that skunk."

"Sure," said Chuck. "The skunk wears a flash pattern, black and white, that says 'avoid,' and all other animals do avoid it—except the domestic dog. I guess that feeling has been bred out of dogs—which is inconvenient sometimes. Maybe you could cross kit foxes with cocker spaniels and come up with dogs that would avoid skunks. You haven't seen Mike Fox's project yet, to see the crosses he's made between coyotes and beagles. They've produced some strange animals: some pups look like beagles with ears that stand up, others are coyotes with drooping ears. He certainly got some bizarre-looking animals. Of course he isn't breeding them for looks, but to study behavior anomalies, hybrid characteristics. Do you know that the coyote has been bred in captivity and is sometimes sold in pet shops? It happens, especially around here, which is too bad, as only an exceptional coyote will remain handleable. Among these crosses of Michael Fox's, some are quite beaglelike in their relationships with man, but others develop coyote characteristics as they get older."

He told me that coyotes, always a topic of interest in the West, do not behave in a typical manner around Tucson. The coyote is by nature a solitary animal; unlike wolf cubs, coyote youngsters tend to leave home as soon as they can fend for themselves. But lately the museum has been getting reports of coyote packs moving around the city. "It's abnormal," said Chuck. "It's because of the influence of man. In the north and east of the town, for instance, people take a great delight in having wild animals come around, and they encourage them by feeding them—even coyotes."

"That's nice," I said.

He shook his head. "I disagree. For one thing it tends to pauperize the animals. They become dependent. They come around in larger and larger numbers, which ultimately leads to a confrontation between animals and humans, in which case it's always the animal that suffers. People call to complain, wanting to know why they are suddenly being belabored by so many coyotes, and invariably we find that somebody near them has been feeding the animals. And they're concentrating the number, you see; it isn't just a question of throwing out the occasional bone, but there are people who buy the best beefsteaks and throw them out to the animals. It causes a breakdown of the normal social organization of coyotes: they begin to travel in packs. There's one group we know about of twelve coyotes—quite abnormal. Another thing: the food they're given isn't good for them. People have the mistaken impression that the best thing you can give any meat eater is good red meat. Well, good raw meat is a horrible diet. It's lacking in vitamins, minerals, and many other important things. Ultimately it can do a lot of damage. In the wild, an animal that has killed gets a sampling of all the stuff he needs; he eats the whole thing. The people who feed animals are well-meaning but misguided. It's a waste of their money, and worse, it makes mischief. We see this so often, not just with coyotes, but javelina, birds—*everybody* feeds birds." He sighed. "It's something we try to discourage as gently as we can, because few people are aware of the danger to bird populations of feeding."

He embarked on a long talk about this, and for a time I thought we were getting too far away from the subject of communication, but then I reflected that the eager if mistaken feeding of all these creatures is, after all, a genuine attempt on the part of human beings to get in touch with them. We are trying to set up a channel.

Hanson was saying, "It isn't written about because the magazines that ought to run that kind of material depend on advertising of bird feeds. Let me give you an example of the dangers of feeding: We have a disease here called trichomoniasis that hits columbiformes such as doves. In the terminal stages of this disease the animal actually dies of starvation and thirst because the esophagus becomes completely blocked with a cheesy substance. It attempts to feed nevertheless. It comes to a feeder where you get a large concentration of birds, and takes up some of the food, which it partially ingests, then drops the infected material, and a healthy bird picks it up and is in turn infected. Then you get a tremendous die-off of doves, finches, and various other birds, all from trichomoniasis, and we are inundated with calls from people wanting to know why they are finding so many dead birds around their feeders. We encourage them to take the dead birds in to the university lab and have them posted: every one proves to be infected with trichomoniasis, but you never see these stories published.

"People naturally think they're doing birds a favor when they feed them, but if a population is very shaky it's going to crash anyway: we aren't going to affect it materially by feeding. You see, by concentrating birds you enhance the possibility of an epidemic. You also enhance predation. Only the other day we had a distress call from an individual who said, 'What am I going to do? There's a hawk killing all my birds!' Well, of course, the person had set up a smorgasbord for the hawk. Without it the hawk would go somewhere else where he'd have to work, but as things are, all he's got to do is sit and wait for some birds to come to the feeder."

"It's not only the hawks," said Doris Ready. "People will buy birdseed and stuff by the bushel, then come in and say, 'We have a problem: we have these little rats running all over the yard. What can we do?' They never put the two things together."

"For another reason feeding's hard on birds," said Hanson. "We have a species of quail here that for part of the year is a grain eater, but in the early spring when there's no grain they eat buds that have a high vitamin content, which stimulates covey breakup, pair formation, and nesting. In areas where they're being fed, they don't take the buds because they don't need food; thus they don't get the necessary vitamins, they don't break up the coveys and nest, and they don't reproduce. People just don't know the damage they do. Take hummingbirds: we have a few around here through December and January. If we have a mild winter they'll make it through, but in a few nights of severe cold the hummers can't ingest enough food before dark, because of their high metabolic rate, to last them until the sun comes up and warms them so they can feed again. So we lose most of these hummers. And they winter here just because people are keeping their feeders out: the hummers just aren't migrating. . . . In short, you gravely endanger the bird population by feeding."

He seemed so downcast at the thought that I was relieved when Mrs. Ready brought the subject back to her kit fox. It had different reactions to different people, she said, and was well aware of the identity of the various people who called at the house. "She likes the humans she's known since she was a little bitty baby, but with new people it's as if she made up her mind on sight. She takes against some, who don't seem to relate very well to animals; sometimes she nips their ankles. But my mother, who as a matter of fact isn't usually very good with animals, is a great favorite with her. You wonder what it is, how it is that animals know which person to like, don't you?"

"It's an intriguing thing," Chuck agreed. "What clues do the higher animals use to relate to a human? As you say, Doris, it isn't necessarily the human's attitude. I'm convinced of that. Animals seem to sense something deeper than a surface attitude. It can't be a sense of smell either, because I've seen animals take a great liking to somebody through a glass door or window. It used to be thought that animals might be turned off by the adrenalin pumping in a frightened or worried human because it produced an olfactory clue that the animal picked up and interpreted as a sign of anger or fear, but you sometimes see an effect, as I said, operating through a pane of glass. One thing we know: The animal makes a decision very quickly, and you can watch it happening. Doris can see it with her kit fox, and I can with my bobcat. I can tell immediately what the bobcat thinks of a person; whether he thinks, 'You're okay, I can relate to you,' or, 'You're somebody I really don't like, and if you come in here I'll show you how much I don't like you.' It's absolutely incredible, and I don't know how it happens."

I told them about an experience of the sort that I had one day at the English zoo Whipsnade, when I was walking there with my husband. Whipsnade has large paddocks to accommodate the hoofed animals, and we were walking past one of these, which was full of various placid grazing beasts, when we saw one animal running toward us. It was a gnu, and its eyes were fixed on my husband Charles as it ran straight into the fence, as if trying to attack him. It recoiled from the collision, but a moment later it was trying again with all its might to get at Charles. It followed us all the way along inside the fence, until it could go no further, showing every sign of enmity.

"What do you suppose it was all about?" I now asked. "To my certain knowledge Charles has never hurt a gnu in his life, and hardly ever goes to zoos. Perhaps something about him reminded the animal of an old enemy—what do you think?"

"No telling," said Hanson. "Take cats. Their etiquette is very intriguing. If you want to get along with a cat you have to know several basic things. Most people when they see a cat—that is, if they like cats—want to pet it and cuddle it. They might say something, too, like, 'Hi there!' Now that isn't the way. You want to look at the animal just for a minute, and after your eyes meet you must turn your head away, to show the cat that you aren't being aggressive. Because the aggressive posture of the cat is the locked-eye gaze and various other attitudes, depending on the subspecies, that mean, 'I am threatening you.' Cats will transfer this reaction to humans, and when the stranger says, 'Hi!' a cat will, according to its nature, back away or make a threatening gesture or merely ignore. There are certain very rigid rules in cat etiquette that you don't violate if you are really going to get acquainted with it. With some cats in captivity this response is, in time, eroded away and they become accepting, but some others never give it up."

I asked if he thought it was true that cats always "know" when they meet people who can't bear them. Chuck didn't think so. He said he had taken such people in to see his bobcat and nothing happened, though on one occasion the bobcat, sensing from the visitor's recoil how he felt, gave him a little push. "It was a testing thing," he said. "He was trying to sense what the man would do. Distressing for the man, of course, but it wasn't important. The cat didn't bite or attack or anything, just pushed, just tested. In other cases he just ignored the people. Yet I've seen times where people weren't quite sure they wanted to meet a bobcat, and he was charming. One would like to know just what is communicated: often I know I'm communicating with him, but I don't know what I'm saying or how I'm saying it. There's no rule I can come up with."

He paused, puffing at his pipe, and thought. Then he said, "No doubt you've heard the generalization that people who aren't liked by other people aren't liked by animals either. Well, there's one individual living here in Tucson who ought to bear that out if anybody does. He's a most unattractive, obnoxious character. And yet I've seen several animals become absolutely delirious when they met him: it was incredible how they loved him. They were positively out of control at the happiness of seeing him. He was pretty matter-of-fact about it: I suppose he's used to it. Up to that point I was smugly predicting who would or wouldn't be liked by animals. I had a dog, an Australian shepherd, that I was sure wouldn't like a certain man because I didn't, but I couldn't have been more mistaken. The minute the dog saw the man he went into fits of joy, proceeding to wet over everything. And he hadn't smelt him, because the first sighting was through a glass door. He didn't hear him either. He just looked up and saw him, and went crazy. Of course it's possible that at some time in the dog's earlier life an individual who looked just like this man had been loving and kind, but I'll never know. One rule, I guess, is that there are no rules."

"Hey, ferret, get out of there," said Doris suddenly. One of the little animals was getting into my handbag on the floor. She grabbed it up and said, "No, no, no, no, no! They're so extrovert," she added apologetically, putting down the unrepentant ferret. "They're into everything, not afraid of anything in the world. We get used to animals running around the office and don't notice them, but visitors are sometimes surprised to come in and find them sitting on our desks. We used to have a screech owl that lived in the bookcase over there. Sometimes if he didn't like a person he'd come down and kick them. One day one of the girls brought in a chiliburrito for lunch, which she meant to eat at her desk. She put it down on the paper and cut it up, and all of a sudden the owl swooped down and grabbed a piece and took it back to the bookcase, and before we knew it he'd swallowed it, cheese and chili and all. Then he sat there with a sort of funny look on his face. We didn't know what to do until somebody decided he must be thirsty, so she filled the bowl with water, and he came right up to it and drank and drank. It all happened so fast, nobody thought it might hurt him to eat such nonowl-type food, but he never showed any bad effects."

I had heard that they never give their animals names at the Desert Museum, and I asked the reason for this. Chuck said, "You see, very few people have any natural history background. If we give a name, people are apt to get all wrapped up in the name rather than the animal's true nature. They're natural pigeonholers anyway, and they tend to be satisfied once they have a name, a handle: that's as far as they'll go. Besides, by giving human names you're arbitrarily assigning human characteristics, and people are far too much inclined to do that anyway. We always explain to children when they ask, 'What's his name?' that it wouldn't be fair to the animal. 'If you call it Sam,' we say, 'you would expect it to do all the things your friend Sam can do, and it can't. It's different from you, and that's what we're going to discuss—the differences.' "

He was silent for a moment, then he continued seriously. "This type of thing, this education, is the only thing that justifies the existence of zoos. There can be no other excuse for keeping animals in concentration camps. I'm criticized for using that phrase, but what else can you call them? Even ours is a camp. We try to make it more pleasant than most, both in our exhibition techniques and in our relations with the animals. We are, I hope, better prison guards, and I think the mental health of the animals is better here than in most. But still, I can honestly say that there isn't a person here on the staff who likes to see these animals in enclosures. We do it because it's the only hope we have of reaching the general public with the vital message about the value of every organism with which we share this planet."

CHAPTER 8

Little Things You Look For

Later I thought about one of the things Chuck Hanson had told me about bobcats and their ways of communication. A bobcat considers it a friendly greeting to be touched just below the base of the tail, where it keeps its spraying mechanism. How in the world would an ordinary human know a thing like that? You would have to be a naturalist, and not only a naturalist but someone—a zoo curator, perhaps—who had to know the animals very well. It follows, I told myself, that if one can collect so much esoteric information in the Living Desert Museum, which is deliberately kept small and restricted, there must be a real wealth of knowledge to be garnered at a larger zoo at which exotic animals are kept. Thinking it over, I asked myself at what large zoo I might find superior people, good at their job not only because they keep the animals healthy but because they use their imaginations and often understand what their charges are trying to communicate. Though I have heard it said that animal people tend to shrink from their fellow humans, I have never noticed it among zoo workers. They may prefer the company of animals, a taste I find easy to understand, but if so they conceal the fact, in the interest of public relations. I went to the Oklahoma City Zoo.

Dr. Lawrence Curtis, the director, made me welcome, as I knew he would, and suggested that we talk first of all to the superintendent, Nick Eberhardt. The superintendent met us in the boardroom behind the office, a tall, thin, graying man with a gentle western voice. His first remark surprised me: when I broached the subject of communication between humans and animals, he said, simply, "Hoofed animals are the stupidest animals there are."

He didn't seem a censorious type generally, so I waited. It soon appeared that his object had not been to condemn; it was just that his attitude was rather like a cowboy's, critical but indulgent and even fond. He went on, "Probably because they're used to a wide range, they don't stabilize very fast in a pen. But I've learned that if you've got a lot of them penned up and just stroll among them, you get them conditioned to you. You hum or whistle and this quiets them, and they associate the feeling with you. When you've done that you can wander around and make any inspection you like, even when they're spooky." He stopped and thought it over. "I think a man can have a rapport with any type of animal. I know I can go in and feel perfectly at ease: there's no feeling on my part of fear, and feeling that they're going to hurt me. I just walk right through, paying no attention, and they stay calm. They can sense if a person's afraid, I don't know how."

Lawrence egged him on: "Does it make a difference, Nick, if you're alone or if you have somebody with you?"

"Makes a great difference," said Nick. "When I've got somebody with me, the animals become more alert: they prick their ears and their eyes flicker and they move to a safe distance, it's a kind of tightening up. If you have to go in with somebody else, to administer a tranquilizer for instance, even if the man with you is experienced, it's better to enter one at a time. And if there's any cover, like a tree, one of you ought to hide behind it a minute until the other's gone around in front. Then it's all right. You can close in and tranquilize your animal or whatever you're there for, and no trouble. Apparently they can't count. But if you go in with a greenhorn you have to be careful, because nine times out of ten you can't convince him not to keep an eye on the animals. He's afraid, and it shows, because he keeps looking around, and they can spot that."

Asked if direct eye contact has an effect on an animal, Nick said that with some animals it does. "In a herd, if you stop short and look straight at one particular animal, for a minute its ears go up and it's alert, and then—it's gone. I can get a reaction from Judy, our elephant, that way, but she only hesitates in whatever she's doing. A direct stare at a gorilla will bother him, but he shows it different ways. Most of them stare hard in another direction, over your shoulder or something, and our big orang, if you look at him, he looks away, too. Funny."

The director said, "Of course, everybody at a zoo takes part in communication of one sort or another with animals, and that's why it's so important to find the right people, people with that quality— a brown thumb, as we call it here. In our job descriptions, when we advertise, we say we want people with an empathy with animals, but finding them isn't easy. Nick's pretty good, though, at picking them out. Nick, what do you look for when you're interviewing an applicant?"

Nick replied with the careful slowness characteristic of him. "It's not a hard and fast rule," he said, "but I've found it's often useful to find somebody who's had experience with domestic animals on a farm, say, or a ranch. That's a plus to begin with, because they're loose when they have to deal with our animals, not all tense and tight like most greenhorns. You don't have to explain to them that there's nothing so dangerous as a stallion, a mule—which is very hardheaded—or a bull. You don't fear these animals, but you treat them with respect, though you don't act afraid of them.

"Usually you take a person you're breaking in on whichever animal and work with him for a few days. A good test is how he'll get along with a big animal like Judy. Generally when I'm breaking in somebody I tell him to chain Judy up. I go in with him and watch how he acts. Judy has never hurt anyone in my presence: I think we have a good understanding. She'll accept anyone I take in. Well, when I tell the man to chain her up, he might try to reach across about six feet to get to her foot, and I know he's no good. You take other people, I tell them how to talk and I introduce them to her, and they go to work right away and talk, and pat her on the shoulder, and she'll behave herself with that person. He'll reach down and chain her and everything. That's the kind of thing I like: I say to myself, 'Well, that boy's got something.' We don't turn him loose with her right away, of course, but in a few days we do. We know her nature after twelve or fifteen years; we know how she'll behave.

"It's just lots of little things you look for when you're trying somebody out. On the other hand, we've had people work here for three days and quit because they couldn't take it. One thing they have to learn: routine in a zoo is an absolute must. You do not break it. We had a lesson in that with a male gemsbok that had been absolutely all right for two and a half years. Each morning when you wanted to go in, the first thing you did was close the door, put the feed out, lean down and wash the rubber water trough out with the brush, clean it, fill the water, open the door, and walk out. This fellow had the same routine every morning for two and a half years. Then one morning one of our men, we'd had him more than a year, he forgot. He went in, put the hay out, washed the water trough and all, but he didn't close the door behind him. The minute the water was turned on the animal lunged out. Directly in front of him was a chain-metal fence about ten feet away; he hit it and broke his neck. It shows how important every step in the process is. You'd think that after two and a half years the animal would be conditioned, but just that one little break was enough to set him off. That's the reason I say that hoofed animals are stupid somewhat."

"Horses, too," I said. "A horse will stand and starve before he'll pull his leg out of a barbed-wire fence."

"Which a mule wouldn't do," said Nick. We agreed, sadly, that horse sense seems to have been bred out of horses.

Nick continued, "For the past few years quite a few people have worked with our elephants. When you get used to them, you can tell how they feel about things. The animal uses his eyes, ears, trunk, feet—oh, you get so you can read him. There are so many things that you look for, and if you learn to know them, you'll never get into trouble. Of course, if you don't watch them . . . Many an employee here has got a slap of the trunk because he didn't know the signs, or he was busy on something else and just didn't look for them. Lots of things are hard to explain to people in words. You've got to learn them for yourself.

"Once, before I knew Judy so well, I started out in a hurry. I was shorthanded that day and not in a very good humor. Before bringing Judy's food in as the man on duty usually did, I went into her quarters and started to clean the place up. I was in a hurry, late, and I had all this work to do, so I says, 'Come on, now, get out of the way,' and I kicked her, and she backed up a little and then she'd come back towards me, and you know, trying to get it all done, I kicked her again and hit her in the shoulder with a scoop, fairly hard, and she just got totally calm. I kept on hitting her, but I thought it was strange her being so quiet. She's always doing something. Finally, I started walking her, and she brought her right front leg out, real slow, and when I turned around she pushed me, broke the shovel right across, and sent me against the wall. Then she just came over and picked me up by the waist and dropped me into the moat, over the rail. I came up out of the moat and tried to get back, but she wouldn't let me. She'd back up a little and I'd try to get out, then she'd come forward again and wouldn't let me out, so finally I had to come out by the front where the moat goes out to the public path.

"I went back around and brought the food in, and she was perfectly quiet. It was all over. But that was clearly my fault. She was thinking, 'Why are you here, late with my food and knocking the hell out of me?' As long as I do what I'm supposed to, she'll tolerate me. She'll do more for a person she likes, but we've got her trained now, or she's got us trained, one or the other. We did have a man here that she was crazy about, though. I never saw anything like it. He would walk all over her, and I think she'd have killed anybody who threatened him. I don't know what it was."

"There's two schools of thought on elephants," said Lawrence Curtis. "In one, you get just one person who becomes completely familiar with the animal, maybe because you haven't got a lot of staff, or it's what the person wants. The animal gets so accustomed to this person that it becomes what we call a 'one-man elephant,' and only that one keeper is completely depended on to manage the animal. In the first situation, it might be all right if the keeper is genuinely attached to the animal and doesn't use his position to blackmail the management, but even then, if he gets sick or goes away, you're stuck with an elephant nobody else can manage. You can have serious problems with a one-man elephant. On the other hand, when you've got a lot of people working with the animal—and that's a unique situation, when you come to think about it, because we work with elephants almost as if they were domesticated animals. In fact, they are semidomesticated: we train them, we tell them lift your foot, put it down, all those things, and if they didn't feel like obeying it would be awkward, they're so big. Are there any other animals we do that with, Nick? I can't think of any, except the big apes in some European zoos. . . . Anyway, when you have a lot of people giving orders, you don't get the chance to develop a real rapport with the elephant. Here, we try to strike a happy medium. We don't want one person to get too attached to the animal, but we don't want too many keepers in on the act, since it only confuses the elephant. We want several, so there can be continuity all the time."

The conversation got around, as it often does, to comparisons between Indian and African elephants, specimens of both being represented at the Oklahoma City Zoo. On direct communication between them and people there seems to be little difference between the species. Nick did not agree with the widely held theory that Africans are untamable, saying that it's a matter of tact. "You can force an Indian to do things, but if you try that approach with an African, you go so far and then see signs that you'd better quit. When our African male gets to balking I generally take a carrot or something and show it to him and say, 'Come on, boy, let's try.' That way you get some distance with him, but you can't keep it up without a rest every so often. You can't push him, and you can't punish him. Tamsy, the female African, works a lot better with kindness. If you give her little treats, she'll actually lay down on her side for you, and that's rare with Africans. Of course, while she's laying there her feet and trunk keep moving; she isn't under perfect control, but she does it, and that's unique.

"Now the rhinos, both white and black, are very different. We've got them sort of semidomesticated, but they aren't very intelligent. It took me half an hour the first time to get them to go out of their sleeping quarters into the pen, with me walking ahead of them and feeding them bits of carrot, and it took a month to teach them the simple routine of the day. We petted them and fed them a lot to get them used to us, and finally they did, but even now they sometimes get upset, running around and snorting when they come out, and I've got to go back and holler at them until they quiet down. It's said they haven't got good eyesight and I guess that's true, because sometimes when we go into the pen to do this or that, tending them, if one of us stands next to a stump they mistake the stump for a man. We still go ahead of them every day, leading them, because we want to keep them used to us in case of necessary medication, et cetera. They're unpredictable. Just the other day one of our young fellows was talking to one of the white rhinos, petting him, when a sparrow flew in and hit against the wall and fell down. Well, they just exploded. The smallest thing can set them off. Yet other times nothing seems to stir them up. Sometimes you can't tell what made them blow up, and some people just don't have the ability to cope when they get that way.

"There's only two of us here who can go in and check for whether any new babies have arrived, because if the mother gets upset at that time she might leave her baby, and nobody else can go in just then. We found out the hard way: once a mother has left the baby she won't go back to it. But our two can go in, like I said, and see what's the trouble in a few minutes, and of course that's the kind of people we want to get. It's the same with the dog animals: you get some people who can go into the pen and stand right close to the animals and never get hurt at all, while others, it's worth their life to go in. Same with the big cats: some people can go in and somehow they respect them and keep their distance. You can go in, do what you have to, and come out. Others better not try. I've never been able to figure out all the reasons for this. I've always had this rapport myself with all types of animals, but I couldn't tell you how it works."

"I've noticed that a lot of people who have endless patience with animals possess remarkably nonabrasive qualities," said Lawrence Curtis. "Their manner and voice are gentle when they're with the animals. But a lot of these same people don't get on very well with other people."

"Which doesn't apply to zoo employees," said Nick rather hastily. "They've got to get along with the public too. We've got a lot of good people here: you can tell by one thing—they never watch the clock. If a job needs finishing and it's the end of the working day, they stay on and never think to complain. Anyway, speaking for myself, my favorite time is early in the morning or late at night. That's when our animals really do their thing, and it's good to be here."

Nick thought back to a traumatic experience that took place in 1963, and told me about it. At that time two veterinarians happened to be working at the zoo; one, Dr. Thomas, was a veteran and the other, Dr. Stringer, was just starting out. The zoo had recently acquired a Siberian tiger weighing seventy to ninety pounds, whose foot had been badly hurt by the leopard, and the two doctors were debating whether to remove the foot or put a cast on it in hopes that it would heal. Dr. Thomas was in favor of amputation; Dr. Stringer was pro-cast.

"I happened to be there while they were discussing it," said Nick, "and right in the middle of it one of the men came to tell me they had a giraffe with a chain caught around its neck. At that time we had pulley doors that were worked manually, and that's where the animal had run into the chain, which had no protective covering. Dr. Thomas and I hurried down and found it had the chain wrapped around three times. You know how these things are: there was no end to the loop, so we couldn't unwind it that way. Well, I went and got a twenty-foot ladder and put it up near the animal. Of course, if it had slipped or something like that it would be good-by giraffe: he'd have broken his neck. Then I got a long stick and approached him with a little flag on it, to get him used to this foreign object before I went any further. Of course, I kept talking to him all the time, with Dr. Thomas standing there watching.

"On three occasions I managed to get two loops off, but every time he moved off the wrong way and got them back again." He laughed. "Finally, after about forty-five minutes of talking and trying, I managed to get all the chain lifted off his neck. I think that if I hadn't had a good rapport, I couldn't have done it. People tend to get too excited dealing with animals, and it scares them. You have to talk soothingly. Anyway, while we were down there getting the giraffe untangled, Dr. Stringer went ahead and casted the tiger's foot." Nick paused to laugh again.

"And that tiger's walking around on his four feet today, perfectly all right," added Curtis.

"Oh well," said Nick judiciously, "two veterinarians, two opinions . . ."

CHAPTER 9

"... An Attention-Getting Device"

At a meeting that afternoon, for which Lawrence had called the keepers together, the conversation about communication with animals seemed naturally to drift, time after time, to the topic of what they called "eye contact"—that is, the messages that are transmitted by human eyes to those of the animals, and vice versa. It is not always a direct eye-to-eye stare, they agreed. Wolves avert their eyes in the face of a direct glance, but that avoidance in itself conveys a meaning, for the direct stare is an act of hostility, which the captive animal does not wish to express. Fred Dittmar, senior keeper of the primates, had a lot to say about eye contact with his charges. Stumptail macaques, sturdy, rather small monkeys, have very expressive faces, he said; he exchanged eye-to-eye glances with them all the time, and had no difficulty interpreting them, whereas to the rhesus macaque a stare into the eyes is considered a hostile move. He always responds belligerently, jumping in place, yawning to threaten the starer, and even taking two or three running steps forward as if about to charge. That is why rhesus in a cage always seem to be bad-tempered when they notice the public looking at them. Colobus, African monkeys with beautiful black-and-white coats, remain indifferent to spectators no matter how long they are looked at, but the brown sakis act as if they were taking a dare when someone looks at them, coming down from their perches and getting as close as possible to the watcher.

"With the orangs and gorillas, there's a lot of eye activity going on," Dittmar said. "On my day off, if I come along behind zoo visitors, they spot me right away: they could find me among a thousand people. The male gorilla and two out of our four females strut when they look at me—that is, they look over my shoulder, never straight at me. That's what gorillas generally do, as if they didn't want our eyes to meet. With the orang you don't get that looking over the shoulder quite so much, but there's one female orang who'll trade things from inside the cage if you offer something to her. If I've got a piece of Brillo, say and there are some others inside the cage, I'll hold it out and say, 'Give me that piece,' and she'll come over, passing two or three other orangs, to reach it and hand it to me. I may say, 'No, not that piece, give me the other over there,' and she'll go get it and swap it for the piece I have. I don't know, of course, if she simply understands my gesture or really hears what I'm saying.

"We've got a lowland gorilla named Fern: whenever I come past the cage she walks over close to me, and I'm just waiting for her to say something. She certainly looks ready to. She walks upright a lot of the time, and looks just like a Neanderthal man. She'll throw things, feces mostly, but if she sees you've got something in your hand, she'll stop and wait. Maybe you don't have anything in your hand, but are just pretending. Well, when that happens she lets fly. But sometimes she'll put her arms up over her head, and I get the feeling she'd *like* me to throw something at her. The trouble is, we're not as good as she is at throwing between the bars. We hit them a lot, but Fern never makes a mistake."

Another primate keeper, Jack Brennan, spoke up with authority: "Fern throws stuff as an attention-getting device more than anything else, whereas the mountain gorilla, Josephine, throws because she's angry; she's hacked off about something. I've heard that when she got here—I wasn't here at that time—she was gentle. I've heard tales of people who could get in with her and no harm done, but it's not like that now. I think what turned her against people, well—in that yard back of the cage where she can exercise out of doors, there are lots of rocks now where there weren't any before. The public's been throwing them, that's what I think. For a while she stopped going out. Another thing: she was real attached to one particular keeper, and when he quit she just went bananas."

I suggested that M'Kubwa, the big male mountain gorilla, might have been beating her up, too. No, said Jack, M'Kubwa never did her much damage. She'd stand up to him and he'd turn away. It was the other mountain gorilla, an elderly lady named Somaili, who suffered somewhat at M'Kubwa's hands—or perhaps she had merely scratched herself on some of the cage equipment.

One of the women keepers testified that she had set up a rapport with Kathy, a lowland gorilla that had recently given birth, simply by staying close to the cage during the first sixty days of the new baby's life. "She would hold it up to the bars and sometimes even present it for grooming where I could reach it through the bars," she said. "Then when the baby got a little bigger she would put it on the floor near me and go off for a little rest. But if I had to go away she would come right back and pick it up again. You might say I baby-sat for her. If the crowd got too much for her, she would come over to where I was sitting and sit down as near as she could get, resting. She knew I was there and wouldn't make any demands on her."

Jack Brennan said he had had three contacts with the baby: "Twice when Kathy left it on the floor I reached in through the bars and picked it up. She came across the cage at top speed, but didn't charge. The third time, the baby got out by itself between the bars. Jim took it off the bars and we carried it around to the back, into an empty cage next door, and put it in again with the others."

"But during the first sixty days I had to stand there," said the woman proudly, "or they wouldn't have gotten close to Kathy at all."

The great apes are not the only animals rendered more communicative by childbirth. Many species, it seems, as long as they are sure of their human friends, become approachable at such times, and some of the infants, starting out in life among human beings, are also more amenable than one might expect. Not that they are by any means tame, even so. The person at the Oklahoma City Zoo who had a lot to say about baby animals was Shirley Kiefer, a big, handsome woman who manages the nursery and looks after such creatures when they are deserted by their mothers. She had evidently had a lot of experience with bears especially.

"I've had brown bears and black bears and grizzlies," she said proudly of her nursery inmates. "You saw that couple out there in the cage just now? Well, they've already bred once, though we didn't expect it. Bears are usually seasonal, and so we didn't think she would produce just yet, but she did have a baby before we separated the couple, and the male got hold of it and killed it. Right now I've got one grizzly cub and one black, and we've just shifted a sun bear out. Bears are interesting, you know: they take a long time to mature. And they do learn to know you. The black bear takes a lot of cuddling and loving, but the grizzly's different. He's just like an adult grizzly right now, mean, you know, even to me, and I raised him. Most animals when you raise them, they stay attached, but he's mean as a bear, as the saying goes."

One of the complaints made by other keepers about nursery animals, said Mrs. Kiefer, is that they've become too familiar with humans for safety: they're not afraid of anything, and will readily charge vehicles and people. One of her babies is a female addax, a kind of antelope, which has long since outgrown the nursery and lives with other adults in a zoo paddock. "They say she gets into more trouble than any other addax they've ever had," said Mrs. Kiefer, with an odd trace of maternal pride. "She just charges—comes right up. Some hoofed stock you just can't tame down, but when they're small newborn babies I take them into the kitchen where I can be with them eight hours a day, and there I can teach them to get used to me and eat. Next I train them to come to me for the bottle, because you can't chase a hoofed animal all over the lot to feed it. Some of those little things get to be just like cows, coming for the bottle. They'll never tame down, but they'll come for their food, to me and sometimes to my assistant. But if a stranger goes in and one of us isn't there, you just might as well forget it. It's always seemed strange to me that a little thing that's never seen anybody but me since it was dropped wouldn't accept me outright as a substitute mother, but the hoofed animals don't. Any other kind of animal, we can do it. We've raised marmosets, orangs, all kinds of primates, cavies, kangaroos, bears—you name it."

I asked Mrs. Kiefer if she thought the tone of her voice was important in her relations with the animals, and she agreed, but, unlike Chuck Hanson, she thought it was, rather, smell that affected them. An animal usually lifts its upper lip to sniff the air at the approach of a human, and that, she believes, is how it detects strangers. "I know they recognize my voice, too," she added, "but in any case they can definitely tell a stranger, every time. There's another funny thing: once in a while an animal takes a dislike to a person when there's absolutely no reason that we can see. I remember an addax that hated me. He wouldn't take food from me, though I never did anything to make him mad, and I worked and worked to make it up with him. I think, too, he knew me by sight, not smell. There's a black panther here—you may have seen her—and she can pick me out of a crowd at any distance within sight of the cage, and I can talk to her and she'll jump up and climb the tree there and roll over like a baby, the way she always did when I was bringing her up. I know that animal remembers, and it isn't a question of smell in her case; it has to be sight. They do remember. We have holding pens here, on the other side of the zoo, and when my animals are big enough to be moved out they're taken there for a while. Sometimes, when it's as much as three or four months after they've been moved, I go over there—because I usually go along with the animals that are being moved; it keeps them from being too upset—and the ones already there know me, and come crowding up to suck my fingers. They really have memory. They go on being babies."

We discussed the opinion of a seasoned animal keeper who maintains that all captive animals are, in a way, in a state of perpetual immaturity, because they are protected and fed and are never in danger. Mrs. Kiefer did not agree. "In some cases, anyway, that's not true," she said. "You take this aardwolf: I've kept him a little more than a year now, and some time ago he started charging the keepers." An aardwolf is an animal something like a hyena. "He didn't attack me, but he did everybody else," said Mrs. Kiefer. "It got so that on my days off nobody could get in to clean his cage. Then, just a couple of weeks ago, he charged me too, and we're having to move him out. That animal hasn't stayed immature."

"Do you greet your tigers?" I asked Wes, who takes care of the big cats. I was speaking of a special noise, a kind of sneeze, that most tigers recognize.

"Yes, I do," he said. "We call that a 'pouf.' They almost always react and do it back again."

"I thought so," I said. "I tried it on a Bengal this morning, and she knew all about it."

"That female's real nice," he said. "The male, though, he just tolerates you, mostly, I don't think he likes anybody."

Mrs. Kiefer said, "It's interesting about tigers and communication. In the nursery, if I'm in the kitchen and hear the noise they make, I can tell what's happening. They have one call when they're hungry and a different one if they're caught in the bar or something, and when a couple of them are fighting it's a different call again. When you hear that tone you'd better go and investigate."

"That's unlike house cats," I commented. "They don't seem to have much variety in their sounds, do they?"

"I wouldn't know," said Mrs. Kiefer. "I've never had a house cat; I don't know anything about them."

"We've got the one Indian elephant and two Africans," Wes said, "and they have different ways of showing anger or fear. The Africans will run up and down sometimes, screaming—not trumpeting, that's a different noise. Most generally, though, if they're angry, they give a low gurgling sound, a slow gurgle, like. The Indian elephant, when she gets mad, just stands perfectly still. The noises she makes are pretty close to the Africans', but you can distinguish them. You can get used to the signs and tell when something's bothering them. The Africans are a little more bullheaded. We've been working with them, and you can tell as the day goes on they get more and more stubborn. They never quite run at you, but it's harder to work."

When a male and a female wild animal are put together with the object of breeding, the good zoo man leaves them together no matter how hostile they may act at first, no matter how frightening their behavior. Often it looks as if one or the other will be severely injured, but one doesn't interfere. You have to sweat it out, I was told. And the ritual is repeated with every breeding season.

"You mean you have to go through this all over again?" I asked. Yes, they said, you have to. It's always good, of course, to give the animals plenty of room for maneuver. But apart from breeding fights, one must keep careful watch of animals together in order to guard against quarrels.

"When those two bears in the nursery play together," said Mrs. Kiefer, "they sometimes get to playing pretty rough. The other day the grizzly bit the black bear in the foot and it really hurt, and the black bear ran over and got behind my legs, whimpering. It was like she had turned into a little bitty cub again, complaining to her mama. Bears are slow compared to some other animals. It takes about ten days to teach a bear to eat with the bottle, where if you've got a tiger in the nursery, after only two or three feedings she'll know just what to do when it comes to the bottle. And leopards, they'll fight you for the bottle right away. They think you're trying to take it away from them: they really fight. A leopard is just about the hardest animal to feed. You have to hold them down until they realize the nipple's in their mouth. I don't think you can ever really tame a leopard down, do you?" The question was aimed at Wes.

"No, I don't believe you can," he agreed. "Cheetahs, now, they tame quickly. Our female, I lead her around on a leash and she's real nice: you can do anything with her. Elephants aren't always predictable. Judy takes likes and dislikes. Some people have been down there feeding her goodies and all that and she pays no attention, won't have anything to do with them. Others can do whatever they like with her. She's just that way. The Africans, anybody can go in with them and they pay no attention, but Judy, if a stranger goes in—we keep her chained, of course—but she'll raise her trunk and go at him and slam him against the wall. Last year she did that, pinned a new man against the wall, and we had a job getting her off him. Yet another new man walked right along with her and she was good as gold."

"I think animals test you," said Mrs. Kiefer. "I was bit once, only once. If you give in and back down, I don't believe you'll ever be able to work with that animal again. You've got to go right back and try it again, that's my opinion. Once there was a ouakari we had to deal with. You know ouakaris, those red-faced, bald-headed monkeys? Well, this one was sick and they put him into isolation. He ran my assistant out of the room, and when I went in he reached out and pulled my hair. So I said, 'Now, just wait a minute,' and he pulled my hair again and charged me. We just *had* to fix him up, you see: he was sick. I wasn't about to let him run me out, so I reached over and cuffed him. He charged me again, and I cuffed him again, and we went on like that until finally I hit him so hard he fell right off the branch, and then we could get to work: we had no more trouble. If I'd given in there wouldn't have been any handling of him at all."

"We had a sick bear once," put in Wes, "he was nursery trained—"

"Not mine, though. I didn't bring him up," put in Mrs. Kiefer.

"Well, he had a difficult temperament," Wes continued. "He'd want to be cuddled, and if you didn't cuddle him he'd fly into a rage, a real tantrum, so you couldn't do a thing with him. I don't think any amount of discipline would do anything with that bear. Nobody could manage him. I sometimes wonder whatever happened to him: I know we sold him."

"Wes knows a lot about communication," Mrs. Kiefer said. "We've had animals born on his day off, and nobody could get the babies out, and they had to call him in, and he never has any trouble." She looked at him proudly.

I asked, "How do you do that? Is it because they know you?"

"Yes, I guess," said the keeper. "Some of them will do things for me they won't for anybody else. The tigers, they'll shift when I tell them to, if maybe they won't for anybody else. Some animals have nice dispositions and some haven't. I know if I take a stranger down with me, the animals absolutely ignore me and try to get at him. I don't know just why."

"I think it's mutual respect," said Mrs. Kiefer, and Wes agreed.

"The rhinos are the same way," he went on. "If you treat them easily and talk gently with them, you'll get along better than if you act real rough. If you're rough they just back up and you can't get any more out of them, you might as well walk out."

They agreed that the sound of a voice is important. Mrs. Kiefer said, "There's two theories on the right way to work with animals. One is that you should let them know by the sound of your voice that you mean it when you give a command—not rough, you know, but firm. That's Wes's way. The other is to be gentle all the time. We used to have a primate keeper who was like that, and all the primates just loved Tommy. It was nice to watch them—but of course primates are different anyway. I think you just have to go with your individual methods."

For a little while we had to break off, because word was brought in that two of the men were needed to help with the loading of an animal that was being sent away. There was a general shifting, collection of coffee cups, and changing of chairs before Director Curtis brought up the subject of feeding times for the animals. How do they always know when it's time to eat? According to Lawrence, it's a twenty-four-hour clock with them. They are expectant even on the two fast days that are observed during the week in Oklahoma City. "They get it right on the dot," he said, "and they come to recognize the footsteps of the man who brings the food. I've seen an animal sitting there waiting, and another guy can go by and the animal won't turn a hair or flicker his eyes, but the minute he hears the meal coming, he's all over the place."

"I know an example: it's the goslings," said another man. "I can drive up to their place in a yellow truck and they don't make much fuss—a little noise, nothing much. But when they spot the white truck that brings the food, they just explode."

"But is it sight or time sense, do you think?" asked a keeper.

"Time, I think," said Lawrence. "You can take the same truck over in the morning, when they're never fed, and the Cape dogs will sit up and take notice a little, but nothing like they do when it comes by between three-thirty and four-thirty. That's feeding time, and they know it. When I was at Fort Worth we had an octopus who was fed every day by a woman. As soon as the octopus heard her footsteps it would start changing color and shape—you know they can be smooth one time and have warts all over the next—but what was really remarkable was the color change. The way it was placed, it could see Alice coming a long time before she rounded the building and arrived. Sometimes we'd get someone else to come around the corner while Alice waited, and the minute the octopus saw it wasn't Alice it would sink to the bottom of the tank again, but it came up again when it finally saw Alice. It would stretch out its tentacles. They're very intelligent animals."

Speaking generally, all the keepers agreed that mutual respect between animal and man was apt to carry dividends. "If you've been teasing and mistreating an animal and then you've got to get into the cage with him, he's probably going to treat you rough in exchange," said one. "On the other hand, if you're on good terms with an animal and you suddenly have to get in with him, I won't say he's bound to leave you alone, but the chances are he will. Look at the time M'Kubwa, the mountain gorilla, wandered out of his cage. When Tommy came and gave him a banana he followed right back into the cage, peaceful as you please."

It was agreed, however, that one can overdo making pets of animals. Such a transformation can lead to suffering. "If you baby them too much it hurts them," said Mrs. Kiefer. "There was a Kodiak bear, not ours, that was babied so much that when they took him away he just moped until he died, and I had a similar experience. I was taking care of a slow loris who wouldn't eat for anybody but me. I had a vacation coming up, and I worked with one of the assistants until we finally got him so he would eat for her, too. Then I went home and started my vacation, and on the fourth day she telephoned and said, 'You'd better come back: the slow loris won't eat.' A few hours later she phoned again and said he'd just died. He simply stopped breathing and died. He was grieving, you see: he was too young to be left."

The talk turned to fear. If a person is afraid, I asked, do the animals sense it as they are thought to do? And if so, can they tell from the smell of the frightened person? The keepers weren't sure. All they could say definitely was that animals reflect moods. If you are angry and upset by something else when you go into a cage, the animal's behavior is likely to be affected—but that is true in any case when one deals with animals: you get back just what you put in. There was also talk about the right degree of relationship a zoo man or woman should have with the charges. Mrs. Kiefer admitted that it was hard not to baby an animal too much. One marmoset came to mind: she had probably kept it too long, and when at last it left the nursery it never did adjust to the change. I recalled that there has been a lot of criticism lately about "humanizing" animals. The cheetahs in the St. Louis zoo, for example, had become very tame and were much attached to their keeper, but they never showed any signs of breeding. Suddenly the word went out that they were overhumanized. Their keeper was told that she must stay out of the enclosure and see the cheetahs only at a distance, when they were fed or otherwise tended. Everyone there was unhappy about it, and the cheetahs showed no signs of becoming interested in each other, even so. New animals had to be procured.

"What we want is for the animals to be independent," said Wes.

"In a zoo?" asked somebody else. There was a pause, broken only when Curtis said it all depended on the animal and what suited it best. He believed the Cape dogs to be happily independent where they were. "But the otters were a different story," he said. "Tell her about them, Ralph."

Ralph said, obediently, "We had three North American otters living down in what used to be the elephant house, all concrete with a very deep moat. There were two females and one male, then we got another male and put one pair down in one of the dog areas. Well, after that nobody saw those otters: they might as well have never been there at all, excepting sometimes you'd catch a glimpse of them early in the morning or late at night. But then the minute they got a look at you they disappeared into their den. Finally we got tired of that and brought them back up here, and within a week they were fine—playing in the water, coming out to look at the public and all. They felt more security up here. Down in the dog area all alone, they felt threatened."

Another keeper said, "There's a situation around the neighborhood where I live, not with wild animals of course, but dogs. There's a mother dog there with one son. They're always seen together; they're inseparable. He follows her and does just what she does. If she feels like going into the back yard next door, he comes running after her. If she runs around the block, so does he. That was all right when he was just a puppy, but he's a big dog now, nearly as big as she is. I think if anything happened to her, he'd die. That's not normal."

We all agreed it was a sick relationship.

"I've been reading this book," I said next day to Rosemary Paces, the supervisor of birds. "The author says something interesting about the way insects, perhaps ants, can tell what time of day it is even when they're locked in a box and can't see the sun. He says we are inclined to take the easy way out to explain this, and give some supernatural reason, whereas very likely it's merely that the ants have a faculty we haven't, and can interpret different polarizations of such light as they can get. This theory doesn't necessarily apply to birds, I should think, or maybe it does. What do you think?"

Rosemary replied, "I don't necessarily feel that we *have* to find an interpretation. I guess I'm lazy up to a point, about what birds are doing and all that, but when it comes to an egg, for instance— you know we have many different species of bird—I find the waiting for it to hatch exciting, that's all. I know what we have to do with it when it's being incubated, turning it over every week and so on, but it's the waiting I enjoy. Our condor has laid two eggs—"

"Yes," I interrupted, "and I've been told you knew in advance that she was going to. How did you know?"

Rosemary thought for a minute. "I don't know how," she said. "I guess it was a combination of signs. He, the male, spent more time away from the nest, that was one thing. And the way she was sitting—you just sense it. You develop a sixth sense. I don't see how any person can spend working day after working day with animals without learning to enjoy it and understand them. Even when they attack you, like a hawk we have. One of the girls who works here— she loves birds, too—was attacked by that hawk every time she went into the Condor Cage." She was referring to a spacious free-flight exhibit, one of the prides of the zoo, inhabited by many species of which the condor pair are the most spectacular. "We tried to figure out why it happened with her and nobody else, and after a lot of talking we found out that she's just simply afraid of him. It didn't make any difference how long she went on and tried to conquer that fear, she couldn't, and the bird knew it. He psyched her out every time. It's always been curious to me how certain people have that fear of large birds, potentially dangerous birds, not the small ones. It happens, I've seen it. The male condor has attacked people two times, people who have worked in other areas in the zoo for years and just dropped in to have a look at the birds. It must be that somebody in his history allowed him to take liberties like that, and he remembers. He's got one particular trick: it always happens after lunch, when the girls are coming back to work. When they walk into the cage, sometimes even when they've gone almost the whole length of it, he swoops down and grabs their purses, and disperses all the contents. We can't break him of it, and the funny thing is it's always at that time of day, after lunch. Though come to think of it, he grabs the public's purses too, anytime. Naturally, people get awfully scared."

"Your toucan likes to swipe Kleenex, too," I said, remembering how the bird once robbed me of my day's supply.

"Yes, I know." She laughed. "But the condors are really characters. They have funny habits. For one thing, they like to take sun baths in the summer, preferably in the middle of the footpath. Last summer at least ten times a day somebody would come to tell me that a big bird was lying dead on the path, and when I got there it was a condor that just reared up and flew away. They're marvelous, totally acclimated. Another of our South American species has taken it on himself to make war on the vultures. That bird feeds his female, bringing her the choicest morsels and helping her to find food, and last year he began to include me in this feeding process. Every now and then I was given this delectable worm or something, and I've never yet brought myself to take it in my mouth—"

"Does he try to put it into your mouth?"

"He certainly does, but I never let him. I get it in my hand, and drop it to the ground when he isn't looking. Recently I reversed the process: I found an insect which I gave to her, and he immediately grabbed it and killed it and I gave it back to her and she killed it all over again and ran away to find him so she could give it to him. . . . And yet, with all this attempted infidelity, they're able to breed and carry on their normal reproductive behavior. She laid an egg last summer. Unfortunately it was taken by some member of the public. Now they've made their nest further away, and we're hoping for better luck. It's so unusual to have that relationship with a bird and yet have him carry on with his normal living pattern. You feel that the bird is proud of what he's doing and very anxious to show it to you. I've had the male come to me on sundry occasions when there's something he's very anxious to show me; he'll peck and fly to something in the corner and look back for me to follow. I think that's very unusual."

Oddly enough, Rosemary was not always a specialist. She started her work in a New York zoo, in her native city, more or less by chance, and her feeling for birds, dormant until then, developed rapidly until she decided that this was what she wanted to do more than anything. Coming out to Oklahoma, she got her chance. "I don't know why it is," she said, "but there's something about birds hatching out. I was thrilled when I saw the birth of number one, and now when I see, say, the thousandth and sixth bird hatching, it's still the same. It's just great. Every time you see that big bird struggling out of the egg, it's neat. I try to instill my appreciation and awe in other people who are watching."

I mentioned the hatching-out noise with which new birds are said to signal their nestmates that it's time to be born. "Yes," said Rosemary, "I meant to talk about that because of what we've been doing with the waterfowl. Unfortunately, our incubator isn't suited to waterfowl, of which we have a wide variety, and we found out that the eggs, which are naturally much moister than those of land birds, tended to dry out. The dry membrane made it much more difficult for the babies to hatch out, so we had to supply humidity by moistening the membrane, perpetually showering it with water. I really think this year I've developed the skill to do that and can save a lot more young birds. Some come out fine, but some have a hard time, and usually they're weak.

"Well, I've found that if I poke them with my finger and talk to them, it helps—just talking. It's really fantastic. Now, I know this sort of thing goes on in nature; when you've got a heap of ostrich eggs all together, for instance, they signal to each other, and the mother's there, too. But I have just one of a kind, and I've found it works to talk. In fact, you really *have* to talk to them to get them out of the shell, and when it's over, you feel you've helped. This is a new thing. My fingers are getting more adapted to chipping away at the shell, and I'm beginning to be able to judge better, when I think, for instance, 'Now shall I help this one out or wait about fifteen minutes longer?' I'm there every morning, first thing. The doors open at seven-thirty and I'm there by a quarter to eight."

I said, "How do you know when an egg's going to hatch?"

"When you've dealt with them a long time, you have a feeling. It's probably something about the balance of the egg. It's in your hand, and even when you can't tell if it's supposed to hatch in twenty-three or twenty-four days, you can feel it, and you know after a while whether you're going to have a successful hatching or not. These waterfowls, they're something new to me. With the big birds it was easier, but so many of these are little eggs, and it's different. You have to learn. Of course I don't prefer incubators to natural nesting, but sometimes you have to have incubating, and it's very satisfying to carry it off.

"Not that the newly hatched birds relate much to the keeper, especially when so many are coming out at once. Penguins, for instance, which look perfectly awful when they're just hatched, team up with each other and start right out with their lives. They're very independent. The bird that's most fun to watch after hatching is the ostrich, because ostriches start running right away, as soon as they're out: they run and run until they just drop. I'd like to watch the development of a generation of ostriches, but we always move them out. They're sweet, but I get awfully mad at them trying to get them to eat. That's always a problem with incubator birds, you see, getting the first one or two to learn to eat well enough to teach the others, and ostriches for some reason are slow. Sometimes it takes as much as two days to get them going."

She paused, then held up her hands, cupped so that I could almost see what she was thinking of. "Just to feel an egg getting ready to hatch," she said. "It's—it's neat." Rosemary is a reserved girl, but her face was suddenly brilliant with a smile.

CHAPTER 10

Actually Talking

"You'll have to see David McKelvey on this project of yours," a zoo friend told me. "Nobody knows more about birds and animals than he does, especially birds: he actually talks to them. I've seen him do it, and it's incredible."

Birds. Yes, I thought, it was high time I paid some attention to these very vocal creatures. Thinking back to the most knowledgeable author I have ever read on the subject, Konrad Lorenz, I smiled as I recalled one of his anecdotes. On a hot day he had taken a friend, a photographer, to a pool in nearby woods where it was hoped that some good pictures might be taken of a brood of goslings he supplied for the purpose. The goslings were not cooperative, and the photographer said crossly:

"Quack, quack, quack—oh, I beg your pardon. Honk, honk, honk . . ."

It was Lorenz, too, who formulated the concept of imprinting and gave it its name. A bird when it hatches out becomes irrevocably attached to the first living, moving creature it sees. In nature, of course, this will be the bird's mother, but nowadays many birds are hatched out in incubators, and the resulting attachment is often grotesquely inappropriate. I have seen films made at Lorenz's farm, where small goslings and ducklings trot along at the heels of the young attendant to which they have firmly attached themselves at the moment of birth. The sight is most impressive.

In Petoskey, Michigan, I found the McKelvey family in an unexpected flurry of packing, for they had just heard that they were
going to Mauritius for two years.

"It suddenly came up," said David McKelvey, a tall, thin man
whose beard did little to make him look older than his thirty-six
years. "We leave here day after tomorrow: the World Wildlife
people and the International Society for the Protection of Birds
have asked me, jointly, to try to save two species of birds out there
that are threatened with extinction. It's the most critical endan-
gered-species program for birds that's ever been undertaken: the
captive propagation and, we hope, breeding of the very small remnant population of the Mauritian kestrel, which is a small falcon
indigenous to the island of Mauritius—it's a beautiful little tan-
and-black-spotted bird that looks very much like what we call a
"sparrow hawk" here, the kestrel of North America—and the pink
Mauritian pigeon. That's a large racing-homer type of bird, a
beautiful strawberry pink with red bill and feet and a cocked bronze
tail and bronze wings and back. And there are only twelve of them
left. Both these birds are threatened by the reduction of the remnant
forest on the island and the introduction of the crab-eating macaque, a species of monkey sacred to the Indian population but
almost entirely destructive of the kestrel and pigeon. We hope to
protect at least one pair of the kestrels, now sitting on fertilized
eggs, and to take into captivity some of the pink pigeons. As both
these birds are at danger level, it's quite a challenge."

McKelvey had been described to me as a man with an incredible
talent for talking with birds and animals, and I had determinedly
tracked him down through the various pet shops, ASPCA offices,
and zoos where he has worked for a time. Now it looked as if I had
nearly lost him again. In the downstairs room of his house his wife,
Linda, and small daughter, Jessica, now took over the task, while he
talked to me, of putting a large number of objects into cases and
trunks. The gentle noise of their talk was nearly drowned out at
times by bursts of song from a small purplish, orange-bellied bird
that hopped and fluttered in a cage that took up almost the entire
wall along the window side of the room. Inside the cage was a tree of
respectable size.

"We're going to have two days' briefing in England, and then we
fly on," McKelvey said. He has a rapid delivery, punctuated now

and then with a gasp for air. "It's quite easy to get to Mauritius nowadays, I understand—just a couple of days' flight—but I can remember when only one boat stopped there in a month. Of course, it makes kind of a rush for us because we weren't expecting it. I've been working with the animal shelter here, and it meant finding somebody in a hurry to take my place, and we had to rent the house too. We're leaving all our large stuff and the birds and animals, of course. They couldn't go to Mauritius."

"What kind of bird is that in the cage?" I asked. "It sounds like a blue jay, but it doesn't look like one."

David said proudly, "Oh, Sam can sing any song he wants. He's a shama thrush from India, a species that's very good at picking up sounds, just like our mockingbirds. In fact, this thrush and the lyrebird of Australia are the best mimics we have in the bird world. Sam'll imitate anything he hears if he likes it. We have a couple of young wild dogs from Southeast Asia, and when they were pups he picked up their sounds so you'd swear there was a dog in the house. Would you like to see the dogs?" He rushed out of the room, through the back door of the house into a maze of barns and outhouses that stood here and there in the snow, while Linda McKelvey and I discussed the chances of schooling for Jessica in Mauritius. Two small dogs came rushing in and greeted me with unusual ebullience, leaping as high as my face, over and over. McKelvey watched them proudly.

"They want you to regurgitate food," he explained. "That's the way they nourish themselves, like wolf cubs. Come on—" and he swept the dogs out again, then returned, carrying a bird. "Here, come over here where we still have somewhere to sit," he said.

I had just settled on the sofa with notebook and pencil when he put down the bird, a vividly colored miniature rooster, on the notebook. It immediately flapped its wings and crowed.

"There," said McKelvey proudly, "isn't he something? That's an Indian jungle fowl, which is to say the ancestor of all the barnyard fowl you've ever known or seen, anywhere in the world. Beautiful, isn't he?"

The rooster crowed and crowed again, then flapped over to a nearby table, where David captured him and carried him out. The room was quiet until the thrush began to sing again, a new song that bubbled and ran like a brook and continued until McKelvey returned.

His interest in natural history began, he said, when as a boy in

Youngstown, Ohio, he spent a lot of time in his grandmother's front yard, a jungly sort of place, full of the flowers and plants she loved. David liked to study the insects and pond life he found there, even though his grandmother was shocked and disgusted by the larvae and snails and slugs he found and boasted about. His grandfather contributed to his training by taking him fishing: once the grandfather helped him rescue and cure and release a crippled baby marten. When David's parents moved to Cleveland, his grandfather presented him with a pair of bantam chickens, and David still has the descendants of those bantams today.

"Grandfather was always the one to help me see what was under a rotten log, or explore a bird's nest," he said. "He cared more about eating animals after killing them, but he helped me to develop. I've always been interested, not so much in what animals eat or where they live, as what they do. Before I knew what it was called, I cared about behavior. I could spend hours and days lying hidden, watching a pair of pigeons building a nest, copulating, raising their young. I was always asking questions, some of which weren't approved of by my mother, about why animals do what they do. I loved to communicate with animals, not from the Alice-in-Wonderland or Bambi standpoint, but by actually eliciting responses from them by mimicking vocal or physical action. I had the limited amount of success any ten- or twelve-year-old child might have, but I seized upon the domestic pigeon as a model for my animal behavior studies, and found them, as I find them to this day, a very challenging, satisfying animal to communicate with.

"As I grew up in Youngstown and Cleveland, to the consternation of my teachers I spent more time in the woods and fields studying the animals than in the classroom or at home doing my schoolwork, and I didn't have many friendships among the other students except for one boy—he's at the Smithsonian now—who shared my interest in animals. He was a keen hunter then, sharing with our contemporaries the macho idea of killing animals, but now he's a keen conservationist, and I feel that in many ways I helped shape his attitude toward animal life."

David had done one year at the university when he was offered a job as gamekeeper, or rather aviculturist, for a millionaire who had a collection of wild waterfowl and was much interested in the conservation of these birds. The work could not have been more agreeable for a person of David's tastes. He was able to study and breed a large number of species such as the paradise sheldrake, the European sheldrake, and the mandarin duck, as well as some of the more common species, including mallards. Though they were not waterfowl, also in the collection were pheasants and "a beautiful horned owl." McKelvey then moved on to another Ohio cultural center, where he was the staff naturalist; a lovely place, he said, with acres of forest trail and a number of animals both local and imported. He learned there to give lectures to visitors, especially school and Audubon groups, on the flora and fauna of the place.

Military service now intervened, but as soon as McKelvey's tour of duty was over he found himself back at work as curator of birds at the Columbus Zoological Gardens, developing, cataloguing, and breeding what he called a very nice collection of birds there until, as too often happens in American zoos, a political upheaval unseated a number of employees, and McKelvey was one of those who suffered. A new mayor, the appointment of a new zoo director—the result was chaos. A man who had hitherto cleaned out the monkey house was related to the new mayor: as a result he was given the job, temporarily, of curator of birds, in order to fatten his pension, while David was demoted to the post of head keeper. While the new curator was there, said David with passion, "he killed my ostriches, made my flamingoes turn from pink to white, broke up a nest of hyacinthine macaws, and failed even to keep a domestic pigeon alive. So I walked out in a fit of disgust. I didn't have the stomach to see it through, nor to see my pride and joy, the bird collection, manhandled. From that zoo I went to the science museum in Cleveland, where I had reasonable success making a collection of owls, and we had a nice lot of waterfowl. I was with them for two and a half, almost three years, and in that time I became somewhat known as a lecturer. I appeared on the 'Today Show' with Hugh Downs about seven times in eight months. I was on the Johnny Carson show, too. I got quite a lot of exposure that way, and got so many speaking engagements that I gave up my work at the Cleveland museum to talk here and there about animal behavior, primarily bird behavior.

"I got awfully tired of trying to find accommodations where I could keep with me a Canada goose, a dingo, and a great horned owl, all in one room, and after more than three years appearing at high schools, sportsmen's clubs, and such, I decided to get out of that line of work. Temporarily I took a job in Naples, Florida, as a trainer of performing birds, Pekin ducks that picked flowers and mallards that played the piano, and peacocks that rang bells and black swans that picked up rubber eggs and put them in a nest—rather hokey, but an interesting example of what operative conditioning [teaching by means of reactions, step by step] can do; it's a simple matter of getting an animal to the point where he wants to listen, and then showing him what he's supposed to do.

"From Florida I went to Elgin, Illinois, as foreman of a national wildlife foundation research game farm, where I was responsible for the production of about sixty-five kinds of game birds every year, most of which were mallards, pheasants, wild turkeys, and bobwhite quail, but we did raise some rare species of geese and other surface-feeding waterfowl, wood ducks and canvasback. That got a great deal of my attention for a two-year period. About that time I was about to have my own first fledgling out of the nest, and there wasn't really too much time to devote to being a family man when the staff of food propagators had been reduced from seven to three, and you were asked to be on twenty-four-hour duty or find another job. Much as I liked my birds, I found another job.

"In fact, the next job was with birds too, at the Sedgwick Studio, in Chicago, a collection of rare birds from around the world. I stayed with Sedgwick for a while. Then again the old spirit of adventure got the better of me. Some rich ad-agency types—I called them rich hippies—decided they would like to develop a commune farm in Wisconsin. They would grubstake me to the farm, and they would use it as a weekend crash pad where they could come and stare groggily into the distance. They had heavy nights of drinking and smoking during those weekends. But I got a nice herd of goats going, a nice lot of dairy goats producing for us, a pretty decent beef herd, we had some hogs and poultry. I got the crop lands all going on half shares with some of the local farmers, gardens going—it really made a nice place to visit, but I didn't want to live there any more because we had an endless supply of long-haired spaced-out types from Chicago, and not only from Chicago: we became a kind of way station for all the hippies in the country on their way to and from California. So we dropped out of that, though I still have a lot of friends among the longhairs, and the rednecks too, if it comes to that."

Sam was singing something new. I wanted to ask what it was, but McKelvey continued on a fresh breath. "From there I went back to the Chicago area, much as I disliked the idea, but that's where the money is—to work for the Noah's Ark Pet Shop, the largest pet store in the country. My job was to take charge of the birds and also keep an eye on the smaller animals, such as reptiles. I had over seventy, usually seventy-five species of birds for sale at all times, ranging from rheas, ostriches, hyacinthine macaws, and fairy bluebirds down to ricebirds and fantail pigeons. The Noah's Ark experience with its tremendous waste of animal life, especially birds and reptiles from abroad, became so overpowering to me that after my usual two-year stint I decided to do some penance for my animal-perishing sins and come to Petoskey as director of the Emmet County Humane Society, where I'm working at this time. And now I'm embarking on the rosy pigeon and Mauritian kestrel. If it works, if we can take them into captivity and get them used to captive feeding, if they nest, we may build them up to a number where the next hurricane or the next group of marauding monkeys won't add them to the list of extinct species. Which brings me pretty well up to date. Now, what did you want to ask me about?"

"Interspecific communication," I said, by which I meant communication between different species.

He repeated the phrase thoughtfully. "Yes, it happens, especially in extraordinary circumstances," he said. "You see, a lot of times the specific releaser mechanisms that normally prevent hybridization will actually forward hybridization possibilities when birds are raised in isolation from their own kind, and even in mixed broods. One of the things the waterfowl propagation people have found years ago is that even though it's easier to take some young Canada geese and young snow geese and young wood ducks and young teal, say, and put them all in one brooder and bring them up together, instead of raising them separately according to species, they may grow up to be perfectly healthy animals, but they don't necessarily identify with the proper species at mating time.

"With isolation you can also do very strange things to the mating processes of a bird. There was a time I had a European graylag goose: I was very busy and there was nowhere else to raise the gosling, so I kept it in a wire cage with a coffee can for food and another coffee can for water. As the bird matured I put food and water into galvanized pails, and added a large galvanized washtub for the bird to swim in. I didn't have much time to spend with him, just saw that he had food and water and a clean cage, and he was always a loner for his first three years.

"When he was three years old and it was time to choose a mate, he ignored the other graylag goose that I had provided him with, a very willing and attractive young female, and decided that he would pair up with his washtub. I took the washtub and got rid of it, and buried a wading pool in his cage, and the only other galvanized lover he could find was the garbage can. He paired up with the garbage can and remained mated to it for three years. He would defend it during the mating season, and run back to it and display in the triumph ceremony, and cackle to it, and when it was time for mating he would climb on it. Whenever the garbage men came to take it out and empty it he would fly at them and attack them until they brought it back.

"Finally we left a note on the garbage can asking the men to carry it off, as it had outlived its usefulness. We got a new one made of plastic, but when the goose saw his garbage can loaded on the truck and flattened in the crusher, he was so irate that he followed the truck down the street for a couple of blocks, until he was killed in the traffic. It goes to show how you can radically change an animal's behavior in early years. He would never react to graylag calls or even to females, but just rattle a piece of metal, and he would immediately act as if there was another graylag calling him in the distance.

"If you raise snow geese with mallards, sooner or later one of the snow ganders will mate with mallards, male *or* female, because once a goose gets into its head that an object of its affections is to be its mate, nothing deflects it. Satisfaction of its sexual desires has nothing to do with it. For six years a male snow goose considered itself mated to a mallard male and followed it around even though the mallard had no use for it and always repulsed its advances, and for that matter had a female mate of its own. It was as if a cord held the snow goose to the mallard. The female mallard had no objection to this snow goose uncle because it defended the mating site with such vigor that no raccoons or crows or rats would dare to disturb it. And the snow goose got the added stimulus every year of helping to raise a brood. When penned for six months with a female snow goose, completely out of sight or hearing of mallard ducks, he got along with her beautifully, but on release he immediately went to the side of his mallard mate of the year before.

"Sentiment and sexual gratification have nothing in common, at least in geese. Jack Miner of the Miner Bird Refuge at Ontario tells of a Canada gander that was mated to a female and had raised young for several years. She was mortally wounded by a volley of shot and fell into a cow pasture, where Miner picked her up and carried her under his arm to a cowshed, where he was going to do an autopsy later. The gander followed along, watching carefully to see what was going to happen to his mate. Miner went into the cowshed and shut him out, then later he himself went out by another door. Later he came back to let out the cow, and the gander was still waiting. He immediately paired up with the cow and remained faithful to her until the cow was sent to the butcher some ten years later. The cow tolerated the gander and even seemed to look for him when he wasn't there, but of course, there was no sexual gratification going on. The gander was very, very protective of the cow every time she calved, and acted just as he would have about a goose when she had goslings. There are even records of an Egyptian goose, which of course isn't really a goose but is more closely related to the sheldrake: it paired up with a human child, and defended that child against the attack of a savage dog even though it meant losing its own life.

"It's interesting to what lengths this imprinting will go. Naturally, young snow geese, say, if they're brought up by their parents and with the brood, all their lives are attached only to other snow geese. A snow goose gander I had refused for two years to mate with a white domestic goose called an Emden—I had no other snow geese around—until I painted her primary feathers black and put a black grin-patch on her belt. Suddenly she was transformed from the ugly cleaning lady into a beautiful goddess, and he took her immediately and remained mated with her even though she molted back into her everyday plumage within a few days. It took that releasing mechanism to remind him that she was indeed a snow goose, at least as far as the markings went.

"I had a pair of swing pouter pigeons that would never mate: he was dark and she was white. The male had a gay relationship with his brother and would pay no attention to her until I took her and washed her with Ivory Snow to take the oil off her feathers, then took some India ink and painted the ends of her feathers black, in the pattern of his brother. He took to her within a few hours then, and they remained together even when she returned to her original color. These animals have instilled in their heads by early training a predetermined image of what their annual or life partners should be. Until quite recently I had a pair of capuchin pigeons, father and son: the father became so enamored of his son that he left his mate to live with him, and they behaved like any mated couple. Of course they didn't have eggs, but when I gave any eggs to them they would incubate, develop pigeon's milk, feed them, and raise them to maturity. They took turns copulating with each other. Other male pairs that I have kept spend an inordinate amount of time building and maintaining their nests: they keep very decorated nests. An ordinary mated pair will just build a nest and put eggs in it, but the male pair doesn't do that, of course, and they keep building the nest, and adding grass, and putting fancy upturns on it, and lining it beautifully. They don't have the specific release mechanism of laying eggs to block that behavior, so they just go on and on.

"Bird breeders know that they have to be careful if they give Gouldian finch eggs to Society finches to raise, because when the infants reach maturity they will try to find mates that look like mum and dad—that is, Society finches, not Gouldian. To pair shaft-tailed finches successfully if they had been raised by Society finches, I had to be careful to keep them in cages out of sight and hearing of their foster parents. If I didn't, they would mate with Society finches every time, and produce offspring that were sterile but very interesting hybrids, exactly halfway between the two species . . ."

He suddenly broke off and addressed the thrush. "Come on, Sam, give us something in the way of imitations, will you? He's feeling very stage frighty today. He tends to stand staring into his food bowl and vocalizing; that's his way of communicating to me that he'd like something more in the way of breakfast. You ask him to say something and he shuts up, just like a lot of conditioned animals." McKelvey laughed.

"Another interesting thing about interspecific communication is that some bird calls mean the same thing to several species. These are alarm calls, mostly. A robin calling in distress because somebody is handling its nest will bring blue jays, Baltimore orioles, and catbirds to help drive away the predator, and that's not altruistic interest on their part, but only because they're attracted by these distress calls: it triggers a defense mechanism in them. Territoriality is immediately forgotten during such stress periods, and several birds from adjoining territories will get together to drive off the cock or cat or snake that is disturbing the nest site, as soon as one parent gives the distress call. This is not exactly mobbing, though it has elements in common: ordinary mobbing takes place when, say, a number of chickadees become aware of, say, a screech owl. They will immediately begin to scold and display and fly around the place where the owl is hiding, making themselves very evident, and they will attract purple finches, nuthatches, woodpeckers, and so on, who will then go through the mobbing rituals and displays with them. But they are more interested in identifying the predator and pointing out to all the birds in the area where he is. That's what mobbing is. It establishes in the consciousness of all the other birds where the predator is at any given time. Here in Michigan when red-winged blackbirds are feeding in a field or flying over a territory, when they sight a sharp-shin or a Cooper's hawk flying around, they will give a—"

Here David made a bird noise, a kind of sorrowful "cheep, cheep," so lifelike that inadvertently I looked over toward the shama thrush's cage. He continued, "This will cause the meadowlarks, the mourning doves, and bobolinks to peer out. Even an imitation of this call will cause them to freeze or bolt for cover. Interestingly, if mourning doves and rock doves are feeding in the same field, when they hear this sound the mourning doves will fly away, but frequently the rock doves will not, because it's not part of their built-in program."

Sam suddenly uttered a cry, which David said was that of a kildeer. "He's going into his routine," he said. "I'll list them as they come out. That's a button quail now. There are some chickadees out at the feeder: can we do a chickadee, Sam? . . . Those are normal shama thrush songs he's singing now . . . that's a European goldfinch, that one. You probably know that one important quality of bird song is in not being heard. The great horned owl, for instance, has a very deep voice." The room was suddenly full of his rendition of the call of the great horned owl. "This is below the range of hearing of most of his prey, flying squirrels and rabbits and dormice, so he can call back and forth in the woods and they can't hear him: they're unable to pick up the vibrations of that low-pitched call. You'll notice in the woods that birds that nest very high in the trees, like the wood warblers and vireos, have a very thin, whispering type of song." He whispered a tiny bird noise. "Little whispering, reedy calls like that wood warbler's. Further down in the trees you find nesting birds like cardinals, grosbeaks, and tanagers, and they have a more melodious, fluting sound. This rule holds true all over the world." He paused to give a beautiful rendition of full-throated bird song, then continued, "And down near the forest floor, among the thrushes—"

Sam interrupted with a burst of noise, and we turned toward the cage to give it our full attention. "He's imitating a budgie now," said McKelvey, "and now it's an Indian chanting shrike. He'll get on to that and sing and sing, and he also does the silly brain-fever bird. That 'tink' sound he just made is a rose-collared lovebird." Sam suddenly burst into a call that needed no interpretation—the comfortable cluck of a domestic hen: "Kut-kut-awkut."

"In the wild, in his own country, Sam would be imitating the local birds and it would be an advantage," explained Sam's owner. "He would imitate them so accurately that they wouldn't know the difference."

The bird song gave way abruptly to a puppy's distressed, hungry yelping, and David laughed. "Bird communication isn't all that auditory," he said. "There's a lot said by posture. If the sky is full of gulls and one of them pulls his wings in out of the flying position and drops his leg, every other gull will immediately start watching him, at least cock an eye at him, and if he starts to descend, the other gulls will try to see what line he's descending by and they'll all descend along the same line and try to get to whatever he's descending to first. You'll also notice a pigeon who's advertising the fact that he's tenant of this particular territory and looking for a mate. When other pigeons are flighting overhead he will hold his wings open to expose the white rump patch and the two black bars on the wings; he's literally flagging down a prospective mate, who looks down and says, 'Wow! That fellow's nicely marked,' and alights there. We find this same signaling behavior in many species.

"The drumming of the jungle cock as he flies to his perch at night is another example. He doesn't crow, but he drums with his wings—" McKelvey made a drumming, tacking sound— "which announces to the other birds that he's going to roost now, and any of his flock that want to can follow him. We find mechanical signaling, too, in the rattling of the wings of the prairie grouse and the drumming of the ruffed grouse. Some birds use parts of their anatomy instead of calls. A snipe, for example, has no song at all, but makes a beautiful winnowing sound by stiffening his tail feathers against the flow of air from his wings. . . . Look at the shama thrush: he's doing a little dance now, like a lyrebird's.

"To get back to the snipe: he will fly back and forth over his territory, turning his body so the wind flows back across his tail, making an artificial sound like—" A stream of rippling, liquid noises came from his lips. "The snipe has a lot of calls that he uses on the ground, but they're rather discordant." He uttered a series of grating cries. "Also the mourning dove can use the primaries on his wings for making, when he wants to, a piping sound. When he wants to be silent he simply holds those feathers close against the others, then he flies very quietly. The goldeneye duck, flying through the fog and mist of open sea, keep together by making a sound with their wings, like that—" a soft whistle—"as they pass by. It's the sound of their wings that keeps them together.

"Even the sound of the ankles of caribou keeps the herd together in snow. As the herd moves the ankles make a clicking sound, just like the ankles of gnus do, and the wildebeest in Africa. Wildebeest instinctively go 'ugh, ugh, ugh' vocally, and their ankles at the same time go 'click, click' as they put their feet down. This keeps the herd together in dust and at night. A lot of these herd animals have to consciously click, making their contact. We'll notice zebra finches getting together, going 'eek, eek,' and even one zebra, kept alone for years, will keep up that contact note. They have to be in a different social situation before they quit.

"You find me at the moment, because of the Mauritian trip, at an all-time, lifetime low in birds. The collection left over now is from a great disbanding of breeding waterfowl and raptor birds, and some mimicking birds. I'm very interested currently in the behavior and behavior modification which goes with mimicking in birds—not your typical parrots and mynah birds, but the Indian shama thrush. I think they're the most accomplished mimics, along with the Australian lyrebirds, that you'll ever run across. Sam can imitate thirty-three species of birds, two species of mammals, and two amphibians. He was not taught to do any of these: it's just stuff he's picked up in his five years of life. And he is able to put two and two together, as far as identification of what he's mimicking is concerned, because when he sees a chicken he cackles like a chicken; he will see a cat or a puppy and meow like a cat or cry like a puppy, and he will see a catbird outside and mew like a catbird. He imitates a kildeer family in distress perfectly. You even show him a picture of a cockateel and he'll imitate a pair of cockateels calling back and forth. I'm not saying that Sam is communicating with me or even with others of his kind, but he is able to make a relationship between an abstract of a picture or the sight of an animal and then he will mimic the sound of it.

"Sam was very confused the first time he saw a Pekingese dog; he thought it was a cat, so he meowed and gave a distress call, and continued to peer at the Pekingese until it barked. Then Sam sat around, fluffed up and looking quite morose, and after that he addressed the Pekingese only by barking. The bird is not imprinted: it was captured as an adult. It relishes the meals I provide, but I've made no effort to train it. In the summer I let it out; we're surrounded, as you see, by cedars and spreading maples, which makes a wonderful wooded area. The bird can use its mimicking powers to advantage, because it will sit in a tree and wait until one of my jungle hens comes along. The hen will be scratching in the ground beneath the tree and unearth a worm. Then the thrush gives the call of a chicken hawk, a long-drawn-out wail like this . . . and the hen will then break for cover, dropping the worm, which the thrush grabs and eats with great relish. Then he flies back to the tree to wait until another hen comes along for him to fool. So those birds put mimicry to use.

"Once he was courting a female catbird, which is quite similar— they fill the same ecological niche here that shama birds fill in India. He would follow her around, alternately doing the shama thrush songs and call notes, until the male catbird approached, when he would then sing full speed like a catbird and fly at him. He evidently thought he could scare the catbird off by expressing himself in catbird vocabulary, but the male catbird always gave him a licking, while the female would look at him aghast, as one would look at a strange Martian floating around. After a week or so of this he would always come back to the house sadder and wiser, ready to behave himself again.

"I had a starling for thirteen years—he was finally killed by the family cat. His name was Reggie, and he learned to pronounce the family name of his species, *Sternus vulgaris vulgaris*. When he saw the cat he would say, 'Kitty, kitty,' interspersing that with the alarm note, 'aaahh, aaahh,' then go back to 'Kitty, kitty, kitty,' thus saying, if you want to interpret it, 'I am alarmed by the cat.' He would also say, if I showed him a mealworm, 'Want a mealworm,

Reg?' after which I threw it to him. Only after he had said it would I throw it, so he learned to speak for his food. Of course I could have taught him to say, 'Would you like a fried egg?' and given him the mealworm just the same; it wouldn't have made any difference to him. Starlings are an excellent source of pet stock because they're free, no laws govern the keeping of them, they'll eat canned or dried dog food, and learn to talk as well as any mynah or African gray parrot. They're common in America now, but originally they were European. Man has taken and introduced species wherever he goes—Australia, New Zealand, North America. *Sternus vulgaris* follows man around, because man has always sort of liked him."

Another example of transplanted animals occurred to McKelvey. He had not been long in Petoskey, he said, when he noticed a number of what he described as an elegant game fowl around a house not far from his, at the edge of the woods. He stopped and asked the man who lived there about them, and the man said they came from an old lumber camp up near Mackinac.

"I asked where he thought they first came from," said David, "and he said they were brought over from England, an old English strain called the Derby game. From the Eastern seaboard they were carried west by travelers to Arkansas. The wagon-train men enjoyed cockfights, and the game chicken could find its own living and produce a few scrawny pullets every year to replenish the supply killed by predators. They found their way north, up here, when the timber was cut; the lumberjacks would bet on the fighting of the roosters. Each had his favorite, and they'd keep it, tied to a log somewhere. They ate the eggs when they could find them, but when they couldn't they just forgot about it. Every fall they'd try to round up the chickens, and have at least a few chicken dinners before they went away. Some of the birds survived in the cedar swamp up here by feeding like shore birds, scratching around in the rivulets that never quite froze over. They'd eat shrimp and little tiny freshwater animals and cress and anything else they could find in the unfrozen water. The air temperature would go down to thirty below, but these imported hardy relatives of the jungle fowl survived here and there, even though predators would kill off the light-colored ones: the ones marked more like jungle fowl or dark partridge weren't taken.

"This old fellow I talked to was going through the swamp when he came on a hen sitting, and a brood of eggs. He brought the eggs and the hen home with him, and that's how he got his start with these birds, about fifty years ago. He keeps them around, eats the extras, and sells a few for fighting. They live up in the pines and the cedars until the cold weather comes, and then he coaxes them into the barn and closes them in. They fly like a partridge and are elegantly formed, unlike the ordinary domestic chickens, which are fat and ugly: slim, trim, beautiful birds. They've existed in this country for years and years, and even in England, before they came to be known as Derby games, they were brought from India by the Crusaders. The Crusaders went to India and found the jungle fowl and brought them back because they enjoyed cockfighting. In other words, the chicken was domesticated because of its fighting ability instead of its eating qualities. Back in England they were crossed with the heavy British fowl, the Sussex and the Southampton and the Plymouth Rock, and produced many lighter birds; the Cornish game has a trace of their blood. So I have in my collection a pair of these birds, tough enough to survive in this climate, where there are bobcats, bears, great horned owls, and what have you."

With a trace of regret David McKelvey listed the many birds he had played host to until the impending Mauritian trip caused him to disperse them. "I don't like to keep them in cages," he explained, "so I turn them loose when their broken wings and so on have mended. I have nothing against the captive holding of birds, but I think it should be done in an area where the bird will feel comfortable." Then he took me to a part of the woods where he knew there were mallards. It was not a deserted, wild place: we stood near the front lawn of a cottage at the edge of the trees, where a rivulet ran through the snowy ground. He gave a loud call, waited a few seconds, repeated it. Suddenly we were besieged by birds, flying low through the air and landing near us, crowding about, looking eagerly and without the slightest trace of fear for the mallard that had called them. We watched for a while how they lined up behind their leader to dive into the water and generally sport around. Then David gave another call, a warning this time, and they fled precipitately, whirring in the air, even scrambling on their feet to get out of the way. In a very short time we stood on deserted land, in the quiet of late afternoon.

Last week I received a letter from Mauritius, written by David's wife, Linda. "I enclose a photo of David's 'communicating' with a pink pigeon," she wrote, and I studied it carefully. In it McKelvey is regarding closely a perching bird in a sort of shelter. The bird's face happens to be turned away from him, but it is clearly at ease, not objecting at all to his close proximity.

CHAPTER 11

"...A Kind of Singsong"

If anyone knew about communication between animals and people, I felt, it would be Dr. Michael W. Fox, of whom I had heard not only in Tucson but in St. Louis, where at the time he was living and holding the position of associate professor of psychology at Washington University. He studied the various animals of the dog family, the canids, at a center in the Missouri countryside. (His name bears a purely fortuitous relation to his favorite animals.)

He has written a number of books on animal behavior and physiology, several of which, such as *Understanding Your Dog* and *Understanding Your Cat,* are popular reading much consulted by pet owners. His *Behavior of Wolves, Dogs and Related Canids,* however, is more technical, and it was in its pages that I found a passage that interested me very much:

The wolf is clearly an acute observer, able to detect subtle nuances in the behavior of its companions, and also of people if it has been raised with human beings from early on in life. This ability is based on the fact that much of wolf communication is non-verbal. Subtle changes in body, ear and tail posture and in facial expression can simultaneously or successively communicate fear and submission, fear and aggression. The eyes of the wolf also communicate a great deal (as in the domesticated dog, but perhaps more so). Subordinates are constantly attentive to the leader, and as soon as eye contact is made, they submissively avert their eyes. This looking away is exaggerated when the wolf is signalling submission. A direct stare is a threat between rivals, alternate looking forwards and away denotes ambivalent feelings of apprehension. . . .

Some people who have raised wolves, writes Dr. Fox, or have at least closely observed them, believe that the animals possess ESP, because a wolf seems to know what the handler is going to do next, and is aware of exactly how the man feels that day. Besides, an observer can see a "subtle rapport" not only between wolf and wolf but between *himself* and wolf, especially when the wolf is accustomed to human company. The animal can pick out his owner from a crowd, and when he sees him coming he will howl and jump around almost like an excited, joyous dog. But Dr. Fox doesn't credit the animals with the faculty of ESP: he explains these phenomena by citing the wolf's talent for acute observation. It's like Clever Hans all over again, he thinks. A wolf can judge many things from the behavior of his handler, even when the man doesn't realize he is showing his thoughts. We humans do a surprising lot of signaling of this sort, the author says, adding, "It has been estimated that approximately 80 percent of human communication is non-verbal, much of which is expressed subconsciously. . . . Man may acquire a wide repertoire of non-verbal expressions as a consequence of learning—through imitation, especially in childhood and early adolescence."

Wolves depend rather on inherited expressions of feeling, even though they learn other things from their elders, such as their position in the pack hierarchy. Nor, says Dr. Fox, is it likely that they have evolved to the point where, like men, they can disguise their true feelings at all times. They are ingenuous: they show their emotions. As the author observes:

Such honesty—a basic morality of communal animals—is surely a prerequisite for social organization. A social organization, not at the impersonal mechanistic level of the ant, but at the psychic level of a more highly evolved social individual. In the wolf society, where each individual contributes to the group and each is more or less dependent on others, we see the emergence of a group identity and allegiance to the pack, co-operation in hunting and in repelling outsiders or rival packs trespassing on their home range.

Wolves have been known to help a mated pair in their pack build their den, and also help to provide food for the young.

It should be obvious from these passages that Dr. Fox approves highly of wolf society and morality, but he is not a wolf himself, and this fact was brought home to him fairly drastically when he was helping to make a documentary film, *The Wolf Man*, for television. He and some other men had just introduced a number of wolves to each other, so that they might photograph their behavior. Eldest among the animals was a four-year-old pair of which the female was in heat, a circumstance calculated to make her mate edgy. The animals were in a long enclosure, one end of which was heavily wooded. They were milling around at the other end, sorting out their dominance pattern, while Dr. Fox, with four cameramen, took up stands in the trees and bush. The four-year-old female kept pushing at her mate with her muzzle, grinning humbly in wolf fashion, following him wherever he went, sometimes actually leaning against him so that they ran in tandem. Suddenly they trotted down along the fence toward the shrubbery.

Even so, the men felt that they were safe as long as they stayed away from the perimeter. The wolves came down to the end of the enclosure, rounded the bush, and started up again, and the men followed, eagerly taking pictures, until the animals unexpectedly reversed their direction. Thus Dr. Fox found himself facing them, head on. Even then he thought it would be all right. They seemed to be looking ahead and paying him no attention, until suddenly the male attacked him. Hampered by his two cameras, one strapped to each wrist, Dr. Fox held his arms in the air and shouted for the handler. Meantime the wolf bit his right hand and arm, then moved on to bite his back and chest, while the female, loyally supporting her mate, attacked his legs. Though Fox was heavily bundled, this hurt, but he had the presence of mind to stop completely still, and whine like a submissive wolf cub. This stopped all activity for the moment, at least. With the wolf snarling in his face, Dr. Fox looked away, avoiding eye contact, and continued to whine. At the same time he tried to move back, but the animals saw this. They immediately attacked again, this time, however, threatening rather than actually biting. Then the handler arrived, upon which the wolves merely threatened Fox the more, until the other man got hold of the male and dragged him away. The female attacked again, but he held out one of the cameras and she concentrated on that, chewing at it until he got hold of her by the neck and gripped hard. Even then, though she stood still, she stared unblinkingly into his eyes, waiting for him to make the next move. He was too shaken and tired, however, to do anything. They stayed locked together until somebody brought a choke collar and dragged her away.

Looking back as a psychologist, Fox felt guilty at having brought on the attack. As he explained, the wolves were not innately savage, but were attached to only one human being, their handler. The male attacked because his mate was in heat, and because, as the younger wolves had already accepted him as their overdog, he felt that his territory was being threatened. Fox had been attacked as a sexual or territorial rival, and the female merely supported her mate. With similarly scientific detachment, he described his wounds. His thick clothing had prevented laceration of the skin, though the wolves' long canine teeth had inflicted deep scratches, but the pressure exerted by the jaws and the side-to-side shaking had resulted in extensive muscle and tendon damage.

"I know now how live prey must feel," he wrote.

A number of photographs were taken during this encounter. "Note 'fear-grimaces' of author and of the wolves' handler," Dr. Fox directs the reader: in the picture both men are indeed grimacing. In his book the author is especially interesting on the subject of facial expression in canids. With illustrations he shows that in wolves, coyotes, and red and gray foxes the cheeks, or sides of the face, are white, and these areas are favorite targets during playful attacks. They also serve as contrast to the animal's black lips, so that other animals can clearly see any changes in expression. When a wolf, coyote, or red fox makes a play face or a submissive grinning one, the lips are pulled back horizontally, or may be pulled forward and retraced at the same time to show the teeth, usually aggressively. Some domesticated dogs do the same thing, but their cheeks are not always white colored, so the effect is somewhat muted.

One often finds dogs grinning at humans, too, and doesn't exactly know what mood they are in. The grin appears aggressive, but dogs don't do it to each other, only to humans, and the expression so clearly resembles a human grin that Dr. Fox suggests mimicry has been incorporated into the dog's repertoire. A black Labrador of my acquaintance had such an exaggerated grin that if one didn't know he was the most amiable of animals, it might give one pause. A nice old lady who had volunteered to look after him over a weekend was quite intimidated, and spent most of her time avoiding the poor animal, though he smiled and smiled as winningly as he could. I thought of him as I read Dr. Fox's book, realizing that he was unlike a number of the animals therein described. For one thing, he wouldn't have resented it if you looked him straight in the eye. He wouldn't have blinked or turned away his face, and certainly he would not have growled or attacked the person staring at him. He was much more like the completely domesticated dogs described by Dr. Fox that roll over when stared at, lifting a hind leg as a sign of submission. He was a far cry from his wild ancestors, but then, aren't we all?

In St. Louis I telephoned Dr. Fox at Washington University and found that he was very busy, sandwiching his lectures between visits to the West Coast and other TV stations. Still, he said, he would be there for the weekend, and if I cared to come out to one of the university's country buildings, where he and his family were chaperoning some undergraduates, we could talk. On Saturday I found the place, a pleasant house in a meadow surrounded by woods. There was also a shed in the clearing with a caged yard attached. I glanced at it in passing, and wondered if it contained one of the coyote-beagle hybrids I had heard of in Tucson, but whatever lived there was staying indoors.

Dr. Fox and his wife, a good-looking pair of young people in blue denim, were outside doing something to their car, but they asked me indoors, where two small children and a couple of cats were behaving as children and cats usually do on Saturday: the children played with each other, sometimes scrapping, and the cats dozed on chair and sofa.

"Animal communication," said Dr. Fox musingly as we settled down. He spoke with an English accent, and I remembered that he comes from England, where he took his first two degrees as a veterinary surgeon. "Why do you suppose there is such an interest nowadays in these studies?" he asked. "Is this an aspect of our creative boredom, in that we have a supportive technology now tending to run away with us? We're after something in life that has meaning. Some people will find fulfillment in art, or in developing new leisure devices, new machines, whatever; others find it in nature. To me the only reality is nature, and nature expresses itself in consciousness through man: ultimate fulfillment in creation is through man. I see the interest today in animal behavior as reflecting man's almost narcissistic curiosity about himself, trying to find out what he is and why he is and where he is going. That can be a red herring, because we simply *are*.

"There's a lot of interest in the West today in Zen and Eastern philosophy and so on—striving for perfection and the higher consciousness—and I think that is all nonsense, too. The simple reality of a leaf, or a puppy dog playing, or a child smiling, is all there is. There's everything there and yet there's nothing, except the personal significance that one gets in an eternal moment of relationship. That's why for myself it's important to understand how animals communicate. Their communication expresses how they feel their emotions. Some people debate whether animals such as dogs and cats have emotions like ours. They have the same emotional centers in the brain and the same basic needs as we have. It's not anthropomorphic to say that they feel and express themselves very much as we do.

"My popular writing in *Understanding Your Dog* and *Understanding Your Cat* is designed to replace sentimentalism and anthropomorphic projections with a deeper understanding which will lead to empathy and the reverence for life. We also have a responsibility now toward all of life. In domesticating animals and plants we have removed them from the natural shaping factors, and we have a responsible position now as stewards of our inbred dogs and our inbred corn and what have you. So another side of our understanding in communication is to understand our role. I get very disturbed when I think about many of the animals on farms being mass-produced to convert one form of vegetable protein into animal protein. What kind of a life do they have in small cages, in small pens?" Dr. Fox stopped, looking worried.

He resumed: "What is a happy animal? What can we learn from animals' behavior that can guide us to more humane ways of keeping them? This leads me now to ask about human values and priorities. There's a horrible utilitarian value that dictates the way in which we regard animals, the way we regard the whole world in terms of exploiting, how we can best use them. That is why in my understanding of animal communication I'm not the least bit interested in claiming them, imposing my way on them, but rather communing with them in a more subjective sense. The wolves that I've raised will respond in different ways to people. With women and children they're very accepting and trustful, but with men they're very wary. They're not aggressive; they're afraid. There's a different way in which men and women relate and interact. The man is assertive and insertive, and goes to the wolf to be friends, while the woman will say, 'Let's be friends together.' And the child before puberty has an energy field which is more diffuse and open to the world, and open to the wolf. You might translate this in terms of the different ways in which the ego is manifested. The male ego, typical of the ancient hunter, is one that probes, asserts, and influences.

"I think part of our kinship with animals comes from our ancient heritage as hunters, in getting an acquired knowledge of animal behavior, knowing their behavior intimately because our survival depended upon it. But now, what is the use of learning about animal behavior? Well, our survival depends upon it in a very different way, because of the integrity of the biosphere: the healthier the environment is, the healthier man is. The natural environment is the reflection of the quality of human life, physically and psychically, I believe. So we must now understand animals in order to detect and preserve as well as perhaps understand that we, as animals, are no less beautiful.

"A lot of the popular animal texts today, such as Morris's *Naked Ape* and Lorenz's *On Aggression* deal with the phallus and aggression, Lorenz going off the deep end about the innateness of aggression. There's no evidence for that in man. Aggression is not a primary drive: it's a survival response in ones frustrated or in conflict, and it doesn't always come out as aggression. I think if we look at animals we see the basic id of man, not being in the Freudian sense the hyperaggressive, hypersexed monster but the animal in man as being very beautiful. In a book I've recently written about relations between animal and man, *The Key to the Kingdom,* I'm attempting to show that man is both animal and God; that there's a biological, evolutionary ecological kinship with all animals, an inseparable, interpenetrated interdependence; that man's now in the position of God, in terms of responsibility and stewardship. It is in this direction, not to control and exploit, but to understand and harmonize, that we are seeking for knowledge in relation to animals, and not the scientific brush-off of knowledge for knowledge's sake, a justification for vivisection sometimes and for all kinds of very inane work in science, where the scientist simply hasn't assessed what he is doing in relation to more global priorities. I see communication in a much broader sphere, many channels, many levels, where we have people in different fields of interest, interest, like different species, where they can't really communicate with each other any more, and there's a breakdown in human society. If we can learn from nature to model society around more integrated, interdependent, interpenetrating lines, we might be able to flow more easily.

"I'm impressed about communication in man in terms of body attitude or posture. You can learn a lot from a person's carriage, the way he sucks in his chest or blows it out, raises his shoulders, holds his head at a particular angle; you can find out very quickly his attitudes, his character if you like, for a lot of these postures are frozen displays of fear, defensiveness, assertiveness, or whatever. It's an old truism that the more aggressive a person is, the more fearful he is. We've found out that some of these areas of the brain, under deep pressure, produce déjà vu memories, as they are called. This is another side gain from studying animal behavior. But back to interspecies communication. It's a fascinating area, and there's a good deal of it in the natural environment, as we found in the jungle of India: different animals know each other's alarm calls." Dr. Fox had recently returned from a field study of the wild dog, or dhole, of India.

"It doesn't have to be a conscious kind of cooperation, but the mobbing of birds, for example—several birds ganging together and mobbing a stoat or an owl—is one example of interspecies cooperation, perhaps. More bizarre ones will develop in captivity. A jackal and a red fox raised together will engage in a lot of grooming. In the fox world the groomee is the one who withdraws, and in the jackal world it is the groomer who withdraws, so when you have a fox and a jackal together and the fox starts grooming the jackal, the fox never knows when to stop. But the jackal is waiting for the fox to stop, so in the end he finishes up with no hair on his eyes." He laughed.

"In the home, a lot of enjoyment can be gained from this kind of multi-pet family, beginning, of course, with the interaction between a cat and a dog. Different signals and similar signals lead to a relationship. Some of the affinity between man and dog, and man and cat, comes from similar ways of communicating. The direct stare is a threat. Looking away and smiling is a friendly gesture of submission. Arm movements in man, pushing away, is an analagous movement in a dog. I like to get people to relate to animals in their own terms, such as blowing into a horse's muzzle, or nibbling it with your hands along its scruff region, or, with a dog, bowing down into a play bound, panting, which is dog laughter, raising an arm like a paw. Some dogs will be completely freaked out by this, and become quite disturbed. Others will suddenly become alive; their eyeballs gleam. . . . You're crossing into another world, and I think this is another aspect of people's interest in animal communication. People like John Lilly . . . space exploring, really, into another consciousness.

"The work with chimpanzees, teaching them the American deaf and dumb sign language, has hit the headlines. I think it's unfortunate that man has not learned the chimpanzee's language first, because one would think man could learn it faster than they could learn ours. The most important thing that has appeared from these chimpanzee studies is that they have the capacity to develop this kind of language and perhaps even more. In the natural environment they were never challenged and didn't have to develop. We don't know the limits of chimpanzee capability—nor do we know the limits of human potential either."

The conversation turned on the rejection of these chimpanzee studies by a number of the nation's serious thinkers. I commented on the fact that they are often almost violent in tone, as if linguists and philosophers have been outraged by the work, fighting the conclusions bitterly every step of the way, giving ground only when proof after proof is presented, and wrangling endlessly over the definitions of language and communication.

"Why do they get angry, do you suppose?" I asked. "Because they do, you know."

"They may be attacking from an intellectual premise," said Dr. Fox, "but there's a lot of feeling behind it. I think that a lot of people are still entrenched in the Judeo-Christian tradition that man is very separate from nature. Man's alienation from nature is implicit and explicit in most Judeo-Christian teachings. Man developed ego and made a natural world out of his ego sphere, as I call it. Animal qualities were seen to be human and therefore were negative—greed, possessiveness, aggression, and the like, but from objective study of animal behavior we know better now. And yet Darwinian theory came along, so we had to accept evolution and the continuity within species. We have to accept some kind of kinship with animals. Still, there was a lot of negativity about accepting that animals have humanlike feelings. We can accept more easily that we have some animallike qualities, as in the book *The Naked*

Ape, but that animals might have humanlike qualities, such as love, altruism, compassion, humor, jealousy, *and* language, is regarded very negatively. I see this again as the Judeo-Christian type of alienation in our thinking, that we are still separate from animals, and this is just anthropomorphic nonsense and speculation.

"For Heaven's sake, you make a section of a dog's brain, and it has the same emotional centers as a preverbal child's. You raise a dog in the same kind of dependent relationship with its parent owners, and you will see psychosomatic and hysterical disorders, well documented by pediatric psychologists, one example being a Pekingese—I always cite it—that was a child substitute for many years. Mother had her own baby, and this was too much for the dog, who developed paralysis in both hind legs, which cleared up as soon as the dog was taken out of the house: it was up on all fours, but as soon as the vet sent it home it collapsed again. There are some examples of high-order behavior, where occasionally, especially in an animal close to you, you will kind of rip open your expectations, like the lady who put her coat on several times, and her dog, anticipating going out with her, was totally confused with this. When at last they did go out, the dog suddenly became lame. She investigated the paw and there was nothing wrong with it. She caught the dog's eye and it was gleaming: the dog was laughing at her and wagging its tail, and then he ran off—nothing wrong with him at all. Constant lessons from nature like this give us caution as well as a sense of humility.

"I'm very concerned about expectations and attitudes. Man tends to relate to animals, nature, in terms of his needs: it's an exploitative relationship. But he also relates in terms of his knowledge (and knowledge gets in the way, too) in terms of attitudes, values, expectations. For example, he has a certain set of expectation values and attitudes toward a cat. He thinks cats do this and do that. They're very solitary, they're aloof, they're not very sociable. So he gets a kitten, perhaps, and he raises it alone with this set of expectations, and it's a self-fulfilling prophecy. It's the same way sexism determines how little boys and little girls grow up in different homes. I'd like a person to be able to see a tree, a stone, a cat, a child, without all this garbage. The human mind has to be secure to classify, to pigeonhole, to set attitudes, and they do get in the way. They stop us from seeing, from getting close, from empathizing.

"We have a negative set of expectations in relation to several animals, the best example being the wolf and all the mythology around it as a vicious killer and so on. My research has helped, as has the research of many people, to dispel this mythos and replace it with the logos of their altruism, their family life, and so forth. I think it's unfortunate that in my children's book, *The Wolf*, which is in many schools, I'm essentially justifying conservation of the wolf because of its fine qualities, the moral qualities being humanlike, while in fact, as with the whales, it is argued that we should save them because they are perhaps more intelligent or conscious than we are. But we should not place a value on any animal: its very existence places a value upon it. So this is negating another motive for understanding animal communication."

Dr. Fox brooded a minute. "Do we really want to communicate with animals, and if so, what is the end game?" he asked. "To control? To enjoy? To commune?" He sighed, and went on, "I'm very concerned today about education. I wrote *The Wolf* after a scientific meeting where everybody was just showing raw data on the slides; there wasn't one picture of an animal, and this was an international ecology conference, going on for ten days. It was a mechanistic approach to animals—no sense of wonder. I asked myself how many had followed Konrad Lorenz's dictum: You must first love your animal before you study it. So I started writing the children's book, because I think education has to be not only the head trip, the human biocomputer; it has to involve the viscera, the entire experiencing organism, the feelings upon which we can place values on what we do, rather than science being without goal—just data, or knowledge for knowledge's sake. I think that is a cop-out, and people should look critically at what they're doing and why."

I said that gathering data is always easier than interpretation.

"And with interpretation you can always be cut down when people disagree with you," said Dr. Fox. "People are simply afraid of making a statement of a point of view. Like my observations in 1960, when I wrote in a veterinarian's journal about paw raising and sympathy involved. A lot of correspondence came in saying, 'Nonsense, animals don't have emotional problems, they're instinctive.' One correspondent could not imagine a dog refusing to use one leg in order to get sympathy: this, he said, was anthropomorphic. Well, I consider it unscientific today *not* to consider these possibilities."

For a time we talked about the word "anthropomorphic." "It's a no-no," said Dr. Fox, "but 'zoomorphic' is acceptable, because it implies that we have some animal tendencies. Darwinism has done that much for us, anyway. You get evidence from the latest writings that animals have very complex communication. It's a here-and-now communication, existential. They communicate basically their intentions, what they're going to do, and they communicate their emotional state. Very often they can communicate more than one emotional state, such as fear and aggression, such as friendliness and fear, at the same time, because they can simultaneously or successively combine various signals, whether it's a visual signal, facial expression—an expression of fear and friendliness can be combined in a submissive grin—or whether it's in a vocal mode, such as a growl-scream, which has fear and threatening."

"Or the dog's barking and wagging his tail at the same time?"

"Yes, exactly. But a stiff tail wagging can be an aggressive signal, meaning watch out!"

He discussed the psychological approach to animals as compared with the ethological. "The ethological way is merely to observe the animal. It's really almost a mystical thing: you'll find it in the teachings of Zen and other philosophies of seeing the thing in itself. This leads to a kind of nondetached appreciation, too. The psychological way, in contrast, uses the animal to evaluate one's hypothesis about something or other, and while this is okay, I think it's quite secondary if one really wishes to understand.

"You mentioned that you were impressed with the affinity animals have for children and vice versa. I think, as I said earlier, that this relates to the kind of energy field a child puts out, because the ego is not so individuated or sex stereotyped culturally. But, also, an animal will respond appropriately to an infant of many different species, because infant animals in general have aggression-inhibiting or care-soliciting behaviors that are very subtle."

I looked at my notes. "Didn't you say, earlier, that you don't think odor is as important for the animal reaction as a lot of people have said?"

"I think odor *is* important for the animal," replied Dr. Fox. "There can be very short-term changes from an animal in the chemical signals it gives. Considering the different wind directions and so on when my wolves are relating in different ways to men, women, and children, odor is not a prime factor, but in animals androgen, the level of male sex hormones, is related to status, and this influences the kind of secretion you get from various sebaceous glands. Now man, unlike most animals, is constantly in heat. He's constantly producing sperm: he's not a seasonal breeder. He's constantly producing androgen, and this can be very intimidating to an animal. How about the odor of fear? Some claim the animal can detect this, but I think he can also pick up twitching movements, avoidance of eye contact when somebody is afraid, withdrawing movements. You can do this with a dog: You can just go up to the dog and say, 'Hi,' and then suddenly draw back. The dog will immediately go forward and bark. I think the visual clues are much more potent to the animal than the chemical clues, which do have a much longer period before they can change. While we can give a visual display and change it within microseconds, it might take several seconds for a particular emotional odor to be emitted. But this we don't know. I wish to hell we did, because the interest most mammals show in odors, in marking, sometimes in marking each other with their scent glands, too, is a remarkable aspect of their behavior. Carnivores have this second scent organ behind their front teeth or under the nasal organ, but we don't.

"One thing that particularly intrigues and excites me is the communication channel of touch. Ashley Montagu has written a book, *Touching*, which reviews the area very well. One thing we discovered in contact with all kinds of animals is that when they are petted the heart rate slows down. This is an automatic nervous system response in the modality of pleasure, and it's an extremely rewarding and relaxing thing for an animal to be petted, which is why a dog will work for a pat almost as well as, or perhaps better than, it would for food. It's the reason why primates will spend eight or more hours a day grooming, and why cats in the house will spend hours a day licking each other. Apparently, too, the one who's grooming also goes through some physiological change of relaxing, so, rather like 'animals that play together stay together,' animals that groom together are consolidating their relationship and maintaining pleasurable proximity. In relation to heart rate, too, we find that the most outgoing wolves and the most outgoing puppies in litters have the highest resting heart rate, indicating that their sympathetic nervous system is set very high. You might think that they would burn out faster, and indeed they may. It might be somewhat analogous to the executive cardiac stress that we see in man, but at least for the wolf and the dog that we've studied, it's a clear indicator of very early temperament that later determines their role in the social group—if you like, their ultimate personality."

He broke off to speak of a book both of us have read and like, *A Kinship with All Life*. In this the author, J. A. Boone, tells of how he was asked to take care of the famous dog actor Strongheart for a while. Though he knew little about dogs when he started out, he found the experience a revelation. Dr. Fox said:

"Remember, in his account of Strongheart he described all the qualities he saw in the dog every day. In his synonyms he could list only good things. In my two animals outside, Benjy the dog and Tiny the wolf, all that I get there is a constant stream of energy, joy, affection, curiosity, playfulness, occasional antagonism when there's some rivalry, but those disputes are usually settled in a well-mannered if not ritualistic way. In fact, it's only man that really suffers deep, unresolved conflicts between what he is and what he thinks he should be. An animal simply is. Dogs will hold grudges, but they usually work it out. I think that's a reflection, again, of man's domesticating influence, though in a wolf pack you will see an omega wolf [i.e., the lowest on the totem pole] that's picked on by the others. Things are not always hunky-dory in nature.

"But generally in nature at least the animals are always very healthy, because only the healthiest will survive, whereas if you go to a typical Indian village—or even here in St. Louis, where we studied three roving dogs—you see very sick specimens indeed. The closer animals get to man, the sicker they get physically and mentally. I think that's another important lesson for us: we have to learn more, understand more, understand the animal's needs, and understand more fully our responsible role as stewards, not as just exploiting husbandrymen or people who have pets simply as play objects for the children."

The practice of breeding dogs for pets, Fox thinks, has led to many bad effects on the animals' physical makeup, or, as he calls it, their phenotype. He had recently attended a conference on the German shepherd, on which he said, "We had a whole day simply on conformation, with a lot of very fine movies on the animals' gait. And most of the dogs had unsound gait. They're disintegrating with domestication, not with inbreeding as such, but with a lack of natural selection for quality control. In domesticating you do destabilize the phenotype, and ultimately the gene pool is disrupted, so you have all kinds of types that would ordinarily not survive in the wild. This is a very sore reminder to us of the horrible position we're in now, that we have to know everything about everything. We're caught on this vicious cycle."

I said, "Wasn't there something said in one of your books about the Saint Bernard and new undesirable developments in its breeding?"

"Yes, it was quoted in a magazine or two," said Dr. Fox. "That's when the problem really surfaced. There's been one hundred, sometimes, two hundred percent increase in cases of Saint Bernards biting people and attacking children—well, one case was due to a pencil inside the dog's ear, and people have to remember that the pet has to be protected from the child to begin with. But once a breed becomes popular, quality control is relaxed, and unstable temperaments will hit the market and ultimately be bred. In the old days animals with unstable temperaments wouldn't be touched. I think, too, a lot of people will get a dog though they're not really in tune with the whole issue of responsible ownership. Anybody with a large dog, especially powerful types like the Rottweiler, Doberman pinscher, or shepherd, should have a special license of competence to handle, and this should be federally instigated. In England there was recently a debate as to whether people should even be allowed to own German shepherds, because a couple of children were killed around Liverpool. Then we have suburban and urban paranoia— people having their dogs attack trained, though they have no knowledge of how to handle an attack-trained animal, plus the fact that many of these attack-training schools are very cruel and use intimidation methods. This whole business of man's social and emotional misalignment with his dog is reaching a very critical point in the United States and many areas of Europe, not only in attack-trained and guard dogs, but in simply free-roaming dogs. Statistics from cities are devastating: sixty thousand unwanted cats and dogs were destroyed last year in Atlanta, fifty thousand in St. Louis. There's something amiss between animal and man."

"Owning German shepherds seems to be a status symbol," I said.

"Status, power for the male ego," said Dr. Fox, nodding. "You need to show power when you're afraid."

He stood up, and we went out, Dr. Fox's wife and children and all, to see Tiny, who it seems lived in the caged area I had noticed before, sharing the space with Benjy the dog. Benjy was not a beagle-coyote hybrid but a plain, endearing mutt. A number of neighborhood dogs had appeared while we were in the house, and had arranged themselves in a large, sparse semicircle, the focus of which was the cage. Michael Fox indicated them with an amused smile, and moved on to open the cage door.

"Come out," he said into the door. "Just you, Tiny. Benjy, stay where you are."

But as Tiny, a beautiful white wolf, emerged frisking, Benjy squeezed out past her and ran at the waiting dogs. There ensued the most awful racket of dog noises—kiyi, woof, kiyi—which evidently left Tiny unmoved. Dr. Fox was busy putting a rope through a loop on her collar, and through the reverberations I heard him saying gently of Benjy, "He's very territorial."

The intruders vanquished, Benjy trotted back to the cage and was shut in, while Dr. Fox manipulated the end of the stout rope tied to the wolf's collar. He spoke to her in a soft voice and she rolled over on her back, playful as a puppy.

"Do you know how to make friends with a wolf?" he asked me. "Touch her there on the loose skin at the groin." I did so, and Tiny twisted her head to look at me, laughing her wolf laugh, tongue hanging out. "There now," said Dr. Fox, "she's friendly now. It's a gesture they use among themselves."

With a lithe twist Tiny came to her feet and we went for a walk with her around the field, uphill and down. She frisked and danced, but made no determined effort to get away from the long rope, and finally settled down to a lope, keeping up with us in a companionable manner. When we got back to the cage she entered it without urging, and Dr. Fox did not hurry to close the door.

"Would you like to hear her sing?" he asked. "I'll show you."

He called the family to help, and they all settled down on the floor of the cage, in a ring around the wolf, leaving her plenty of room. Then they threw back their heads and began to howl: "Aooo, aooo, aooo." I joined in, until the noise we made was loud and, I thought, rather musical. Tiny thought so, too. For a little while she paced around and around the circle until, softly at first with premonitory whimper, she, too, threw back her head and sang as she paced: "Aooo, aooo, aooo . . ." At last all our noise ceased to quiver on the air, and died away.

"It's social," said Dr. Fox, brushing off his clothes as he came to his feet. "It's a kind of singsong."

CHAPTER 12

Neam Chimpsky & Co.

In 1976 a little book, *The Question of Animal Awareness,* surprised a number of readers not so much by its content as because it was written by Donald R. Griffin, professor of animal behavior at Rockefeller University, a hardheaded scientist who has made a name for his work in various fields but especially on echolocation in bats. The nature of the book was what surprised them. In his preface Dr. Griffin explains that he wrote it because of the wealth of scientific discovery that has lately occurred, or, as he puts it, "a ferment of constructive excitement [that] is evident in ethology." Ethologists, he says, are now confidently making statements about the natural world that differ qualitatively from anything scientifically thinkable forty or fifty years ago, and an important number of these statements have to do with communication, or signaling behavior. Among recent discoveries Griffin lists the gestural communication between chimpanzees and human experimenters, "widely recognized as a breakthrough in the behavioral sciences," as he describes it, but other systems of communication in other species have also emerged: those of fiddler crabs, honeybees, spiders, and leaf-cutting ants, to name a few. There are fishes, for example, that use electrical orientation, and communicate by electrical signaling. Fireflies exchange light flashes. At the end of his book Dr. Griffin makes the suggestion that scientists should construct models of these various species so that more communication might be set up, helping us to learn what is now hidden from us.

The gestural communication between chimp and man has caught the public imagination most readily, no doubt because chimpanzees are so like us that we can think of them without too much effort as almost-humans or even children: it does not strain the imagination to picture a back-and-forth exchange with them. However, these experiments have been bitterly assailed by some people who are touchy on definitions of speech, language, signaling, or whatever. There was at first strong resistance to putting the name of speech to the honeybee "dances." One might wonder why, as what the honeybee is doing is imparting information, but, as Dr. Griffin says, its communication system "edges so close to human speech in its symbolism and flexibility" that a lot of people, whether or not they know why, simply can't accept the implication that man is not unique. For a time the scientist Adrian Wenner, who studied von Frisch's work, and a number of his colleagues even doubted that bees really convey information with their dances as to the locality of food or a good nesting place: it was only after experimenters had devised a great number of tests to prove the contention that they relinquished their objections.

Why are such skeptics so vigorous in declining to accept certain conclusions, so that they must be convinced and convinced further? Because, Dr. Griffin suspects, they are emotionally involved, believing that human speech is peculiarly distinctive, and this distinction "is held to be *the* primary difference that distinguishes human beings from animals." Many philosophers and linguists argue that human speech is closely linked with thinking, if indeed it is not identical and inseparable. So it follows, they reason—perhaps subconsciously—that animals, being animals as they are, can neither think nor speak.

"Language is an expression of man's very nature and his basic capacity," wrote the neurologist K. Goldstein twenty years ago in an article on the nature of language in *Language: An Inquiry into Its Meaning and Function,* edited by Ruth Nanda Anshen. "Animals cannot have language because they lack this capacity. If they had it, they would no longer be animals." He added, in a letter to his editor on the subject, "Man *is* language." Such flat statements are always risky, but people, especially when they are linguists, seem to go on making them.

There is Noam Chomsky, for instance. He writes a good deal about the nature of language, and shows great originality in adhering to the ideas of Descartes, who has otherwise been out of fashion for many years. It was inevitable that Chomsky should have set himself against any notion that animal communication resembles ours, since he believes that humans carry in their minds the pattern of language: that grammar in all its complexity is innate. Like Descartes, as Griffin says, Chomsky implies, if he doesn't assert outright, that animals are machines, and he expresses Descartes's ideas in his own words, thus: ". . . man has a species-specific capacity, a unique type of intellectual organization which cannot be attributed to peripheral organs or related to general intelligence." A few sentences later, paraphrasing his mentor, he says:

Human reason, in fact, is a universal instrument which can serve for all contingencies, whereas the organs of an animal or machine have need of some special adaptation for any particular action . . . no brute [is] so perfect that it has made use of a sign to inform other animals of something which had no relation to their passions . . . for the word is the sole sign and the only certain mark of the presence of thought hidden and wrapped up in the body; now all men . . . make use of signs, whereas the brutes never do anything of the kind; which may be taken for the true distinction between man and brute.

Descartes, of course, lived a long time ago, and there have been discoveries made since then, especially on the chimpanzees and their sign language, to disprove these statements, but Chomsky remains inflexible. Some years ago, in his book *Cartesian Linguistics*, he said:

The unboundedness of human speech, as an expression of limitless thought, is an entirely different matter [from animal communication], because of the freedom from stimulus control and the appropriateness to new situations. Modern studies of animal communication so far offer no counterevidence to the Cartesian assumption that human language is based on an entirely different principle. Each known animal communication system either consists of a fixed number of symbols, each associated with a specific range of eliciting conditions or internal states, or a fixed number of "linguistic dimensions," each associated with a non-linguistic dimension.

Chomsky's comments were published in 1966, but if he has thought better of them he has yet to say so, and the Washoe sign-language experiments began back in 1969. Chomsky was not, however, representative of all observers even before that eventful year. Plenty of people before his day had recounted incidents among animals that led to the assumption of thinking faculties, and not all of them were fond little old ladies who were sure their cats could think because they seemed so clever. One of the more respectable of these observers was George J. Romanes, zoological secretary of the Linnean Society of London in the latter part of the nineteenth century. Romanes was an enthusiastic admirer of Darwin. In 1881 he published *Animal Intelligence*, a collection of anecdotal data on the subject, and he enlarged on the material in 1883 with his book *Mental Evolution in Animals*, in which he published posthumously an essay on instinct by Darwin, which ends:

It may not be logical, but to my imagination, it is far more satisfactory to look at the young cuckoo ejecting its foster-brothers, ants making slaves, the larvae of the Ichneumonidae feeding within the live bodies of their prey, cats playing with mice, otters and cormorants with living fish, not as instincts specially given by the Creator, but as very small parts of one general law leading to the advancement of all organic bodies—Multiply, Vary, let the strongest Live and the weakest Die.

Romanes's collection of anecdotes is shaped to show that animals do think and reason. Though none of his examples strains credulity, before Darwin he could hardly have got away with any of them, and he fully realized it. How much of the resistance met by such explorers into animal cognition is covertly related, as Dr. Michael Fox believes, to religious scruple? Probably a good deal, though it is a moot point which looms larger, this or the conviction that we have been, are, and will always be superior to all other forms of life in the universe. There are still people, though their number is dwindling, who refuse to take seriously the work being done with Washoe and other chimpanzees to determine their capacity to communicate in human fashion.

These experiments were started by Dr. R. Allen Gardner and his wife, Dr. Beatrice T. Gardner, of the University of Nevada at Reno. As they wrote in one of their papers on the subject, it was already well known that chimpanzees could learn to make different responses for different goods and services: what was not well known was to what extent a chimp could acquire a bona-fide human language. Fortunately, the Gardners did not wait until linguists and psychologists had come to terms and produced a satisfactory definition as to just what language is—a definition that has not yet been evolved—since, if they had done so, Project Washoe would never have got off the ground. Instead, they simply started.

They reasoned that any theoretical criteria that can be applied to the early utterances of human children can also be applied to the early utterances of chimps. If children can be said to have acquired language on the basis of their early performance, and if the chimps' performance matches that of the children, then the chimps, too, could be said to have acquired language. Their chief difficulty was that chimpanzees seem constitutionally incapable of utterance— apart, that is, from a few natural sounds—so the Gardners decided to teach their newly acquired chimpanzee, Washoe, an infant female about eight months old, the American Sign Language (Ameslan) used for some deaf and dumb children in this country.

They reared Washoe with a routine as like as possible to that applied to rearing human children, although all her companions communicated, with her and with each other, exclusively in Ameslan. Within fifty-one months of her arrival, the chimpanzee was using 132 Ameslan signs to express herself and could recognize many more that were signed to her. Since this early work much more of a similar nature has been done, many more chimps are being taught the sign language, and Washoe herself has been moved to a chimpanzee colony at Norman, Oklahoma, where some of the other animals have been taught and become reasonably proficient in the signs.

Of all these chimps, one named Lucy is probably the most advanced. Raised alone by her surrogate parents, Jane and Maurice Temerlin, as a human infant and never seeing another chimpanzee until quite recently, Lucy advanced at an amazing rate. By the time Washoe was using strings of words, Lucy was naming objects with labels depending on her own observations. As if to refute Chomsky's statement about "a fixed number of symbols" and so on, she has dubbed a duck "water bird" and given two names for watermelon, "candy drink" and "drink fruit." Having bitten into an old radish, she called it "cry hurt fruit." Here and there, in various parts of the country and usually at universities, other people are teaching other apes to use the sign language. In San Francisco a young gorilla, Coco, is learning her lessons from Ms. Penny Patterson, and I have seen a very young orang in Oklahoma City get so far as to ask for a drink by sign. One chimpanzee, under the management of Dr. H. S. Terrace, is getting his education at Columbia University: though he is called Nim for short, his full name is Neam Chimpsky.

But Ameslan and the Gardner method are only one branch of the new work. Another method of imparting language, or at any rate a means of communication, to a chimpanzee was devised by Dr. David Premack, then of the University of California, who has now moved to Pennsylvania. Working with an animal named Sarah, who was six years old at the time, Premack used plastic pieces of differing colors and shapes, each backed by metal so that it would stick to a magnetized metal board and could be moved around in a vertical plane. These chips represented words—to which they bore no relation in shape or color—which Sarah learned. In time she also learned some of the rules of syntactical arrangement. She understood how to ask questions, how to say "no" and "not," and even the conditional "if . . . then."

Roger Brown, the linguist, criticized the Sarah project because, he said, it is impossible to know if she really comprehended what she was spelling out with her plastic chips. Also, he claimed, not enough care was taken to avoid the Clever Hans syndrome. However, at a symposium on language and communication held in 1975, Roger Fouts, who has worked for a long time with Washoe and other chimps in Reno and Norman, came forward in defense of Sarah's accomplishments. Quoting Wolfgang Köhler (The Mentality of Apes), who worked with a colony of chimpanzees in Teneriffe early this century, he said, "The decisive explanations for the understanding of apes frequently arise from quite unforeseen types of behavior, for example, use of tools by the animals in ways very different from human beings. If we arrange all conditions in such a way that, so far as possible, the ape can only show the kinds of behavior in which we are interested in advance, or else nothing essential at all, then it will become less likely that the animal does the unexpected and thus teaches the observer something."

It may be, therefore, said Fouts, that one can criticize Premack for arranging conditions in advance so that Sarah could show only the types of behavior that he was interested in, but Premack made it quite clear from the start that he was interested only in certain behaviors, so perhaps he should not be criticized for having done exactly what he set out to do. Roger Fouts admitted, however, that the chance of a Clever Hans reaction had not been obviated in Sarah's case, whereas with Washoe and her classmates due precautions are always taken.

The third method so far worked out to make possible two-way

communication between man and chimpanzee has been developed in Georgia, at the Yerkes Regional Primate Center. Dr. Duane Rumbaugh of Georgia State University and two colleagues, Timothy V. Gill and Ernst von Glasersfeld, use a computer with two consoles and keyboards containing twenty-five keys. One of these consoles is used by Lana, a female chimpanzee who lives in a room surrounded by transparent walls; the other is operated by a human in the larger room outside Lana's living space. The design of the computer depends on a language known as Yerkish because it was created at the Yerkes Center. Each key carries a lexigram, a combination of design elements and colors.

As von Glasersfeld explains it in the book *Language Learning by a Chimpanzee*, the first task was to design elements that could readily be distinguished from one another, could be superimposed on one another, and would yield, after the superimposition, combinations that were discriminable. Nine elements used either singly or in combinations of up to four would yield 225 different lexigrams, and the scientists decided that these were more than enough, especially when they were combined with one of three primary colors, three intermediary colors, and the black that resulted from the use of no color at all. Seven colors added to the design elements would make 1,785 possible combinations—far more than necessary in a language suited to a chimpanzee. Yerkish lexigrams have one meaning each, which in most cases corresponds to one meaning of an English word. Various arrangements have also made it possible for Lana to use the conjunction "and," and there are various so-called vending devices with which Lana on her own can turn on music, look at slides, and even open the window.

A teletype outside the chimpanzee's room prints everything that passes between Lana and the operator, or even whatever Lana may say when she is alone. Thus in the morning there is a record of what she has typed. (The best-known of these night thoughts is probably the message found in the morning that said PLEASE MACHINE COME PLAY WITH LANA.)

There comes a time when one wants to see more than what the newspaper story tells the public about such matters, so I was keenly interested to hear, through a friend at the National Institutes of Health, that America's leading primatologists intended to convene in Atlanta in October 1976 at what was called the Robert M.

Yerkes Centennial Conference, to discuss Lana and other recent developments in the study of chimpanzee communication. The NIH came into it because it is in part responsible for having founded the Yerkes Primate Center and six similar centers in the United States, so my friend was in close touch. Through him I obtained permission to attend the conference, and duly made reservations at the host hotel.

At the morning session there were a lot of notable figures in the field, including the son and daughter of the late Robert M. Yerkes, with whom it all began. Dr. Yerkes, who died in 1956, was the most important primatologist of his time and perhaps of this as well. In the words of one of his colleagues, "His original work with animals antedated the rise of behaviorism, took Gestalt psychology in its stride, and remained throughout in the broad evolutionary, physiological and functional tradition that he called comparative psychobiology." He taught at Harvard until, in 1917, he left to take part in war work, and he was still between universities when, in 1923, he acquired a pair of young chimpanzees. Until then these animals had been seen chiefly as entertainers on the vaudeville stage or in circuses, but Yerkes, who kept them on his farm in New Hampshire, studied them as a psychobiologist and found them fascinating. When he took up a new chair at Yale the animals went along, and in the course of time he acquired five more chimps.

After some time in New Haven he began to wonder if his work with the apes might not be better done somewhere in a milder climate in the country, and he managed at last to amass money to found the Yale Laboratories of Primate Biology at Orange Park, Florida. The institution opened in 1930 with Dr. Yerkes as director. It became famous among anthropologists and psychologists, and when Yerkes retired from the directorship in 1941 his name was retained and the institution became the Yerkes Laboratories of Primate Biology.

In the fifties, when fiscal support grew difficult and Yale found the burden too great to bear, Emory University announced its willingness to take over the responsibility. The arrangement was accepted, especially as the NIH had signified its willingness to help turn it into a new primate center, and all the animals and movable plant were taken across the state line to Emory, outside Atlanta. Thanks to a grant from the National Heart Institute, an imposing building was erected at a safe distance from the campus. It is surrounded by wooded land and has every convenience, but for old-timers it doesn't compare with the amiably easy and makeshift arrangements of Orange Park.

That morning there was a general atmosphere of homecoming week, with people joyfully greeting one another as if they hadn't met since the last conference, which was probably true, and the first session was given over to reminiscent talks about Dr. Yerkes's tutelage and leadership at Yale and Orange Park. Yerkes's daughter, Mrs. Roberta Blanshard, and his son, David, started the ball rolling, followed by a number of speakers well known in the related fields of anthropology, primatology, and psychology: Dr. Meredith P. Crawford, Dr. George Haslerud, Dr. Harold Coolidge, and so on. Dr. Crawford talked of his five years at New Haven and Orange Park, Dr. Haslerud reminisced about Orange Park in the mid-1930s, Dr. Myrtle McGraw showed a film she made in those days entitled "Mother Chimpanzees with Their Young Infants," followed by more reminiscences by Dr. J. H. Elder, Dr. S. D. S. Spragg, Dr. Harold Coolidge ("Happy Memories of My Friend, Robert M. Yerkes, and Prince Chim"), and Dr. Vincent Nowlis. The only exception to this group of people with similar interests was Dr. Karl Pribram, a neurosurgeon, who spoke on "What Monkey and Ape Can Tell Us About Human Language." Then it was lunchtime and more reminiscences.

CHAPTER 13

The Name of That Is "Book"

Dr. Duane Rumbaugh, Lana's headmaster, was chairman for the afternoon and also the first speaker, his topic "Ape Language Projects: A Perspective." He explained that the studies were aimed at discovering whether or not the type of communication characteristic of man, that is, open and without limit, is species specific, a uniquely human attribute. It is probably impossible to identify exactly, said Rumbaugh, the origin of the idea that some chimpanzees and other great apes might be taught to communicate, but Dr. Gordon Winant Hewes, professor of anthropology at the University of Colorado, would shortly discuss this matter at length. He mentioned, however, one of Dr. Hewes's comments in advance: concerning the eighteenth-century philosopher de Lamettrie. In 1748 de Lamettrie announced that he thought apes might be capable of learning language.

It was nearly two centuries later that Yerkes, too, began speculating that apes might have capabilities that would allow for the mastery of at least limited linguistic skills. Though he wasn't very optimistic about the possibility of a chimpanzee actually learning to speak, he didn't absolutely disallow it. In 1925 Yerkes wrote that although the young chimpanzee uses significant sounds in considerable number and variety, it does not, in the ordinary and proper meaning of the term, speak. Consequently, he felt, there is no chimpanzee language, although to his mind there certainly was a useful substitute that might readily be developed or transformed into a true language if the animals could be induced to imitate sounds persistently. With his associate, Blanche Learned, Yerkes did try for a time to persuade his cleverest chimp, Chim (or Prince Chim, as he sometimes called him because of his genius), to speak by doling out pieces of banana to the ape and saying at the same time, "Ba," so that Chim would associate the noise with the gift of this preferred fruit and might thus be tempted to get more banana pieces by making the same noise. But it didn't work; after two weeks of twice-daily lessons, both Chim and the experimenter lost interest. Other, similar efforts were likewise abandoned, since the animals showed no tendency to imitate, and in the book *The Great Apes*, which he wrote with his wife, Ada Watterson Yerkes, Yerkes at last concluded:

Evidently, despite possession of a vocal mechanism which closely resembles the human's, and a tendency to produce sounds which vary greatly in quality and intensity, the chimpanzee has surprisingly little tendency to reproduce other of the sounds which it hears than those characteristic to the species, and very limited ability to use these sounds either affectively or ideationally. . . . Everything seems to indicate that their vocalizations do not constitute true language. Apparently the sounds are primarily innate emotional expressions. This is surprising in view of the evidence that they have ideas, and, on occasion, act with insight. We may not safely assume that they have nothing but feelings to express, or even that their wordlike sounds always lack ideational meaning. Perhaps the reason of the apes' failure to develop speech is the absence of a tendency to imitate sounds. Seeing strongly stimulates to imitation, but hearing seems to have no such effect. I am inclined to conclude from the various evidences that the great apes have plenty to talk about, but no gift for the use of sounds to represent individual, as contrasted with racial, feelings or ideas. Perhaps they can be taught to use their fingers, somewhat as the deaf and dumb persons do, and thus help to acquire a simple, non-vocal, signed language. . . . But speechlessness notwithstanding, intercommunication is highly complex and useful in the chimpanzee. . . . According to valuable and reliable data, the chimpanzee is much closer psychobiologically to man than is the gibbon, and closer than is the orangutan. . . . Indeed, were he capable of speech and amenable to domestication, this remarkable primate might quickly come into competition with manual labor in human industry.

There were various attempts made early in the century by other experimenters to teach chimpanzees and one orangutan to talk, but in spite of the most arduous efforts their subjects mastered only one or two crude approximations of words. More recently the couple W. N. and I. A. Kellogg, with their chimpanzee, Gua, demonstrated that the chimp does have a rather well-developed ability to understand simple vocal commands, to sixty-eight of which Gua was able to respond appropriately. She seemed to have at least a receptive competence for language, if not an expressive one.

Keith and Cathy Hayes were the first to develop a prolonged program, from 1947 to 1954, with a chimpanzee, Viki, that included an attempt to teach it to communicate via speech: Viki was kept in their home and subjected to a wide variety of projects, yet efforts to teach her speech met with only a modicum of success. She mastered four voiceless approximations: "Mama," "Papa," "cup," and "up." She clicked her teeth as a sign that she wanted to go out in the car, uttered a "tsk" when she wanted a cigarette, and had another sound, said to be reminiscent of one made in certain European languages, as a sign that she wanted to go outdoors. Ironically, the Hayeses' failure led them to consider the use of a gesture language—in part because of a written suggestion by Gordon Hewes. However, they did not attempt to teach Viki a gestural language, though their reports are full of instances where she used gestures on her own initiative to attempt communication: for instance, she would ask for a ride by gesturing to the car outside, by bringing a purse always taken along on car rides, by showing a picture of a car to the Hayeses, or by clicking her teeth as a "word" for the request.

Since then a great deal has been learned about the anatomic restrictions of a chimpanzee's vocal tract, and we know better why the animals cannot produce human sounds. If this had been understood in 1947 when the study with Viki began, the Hayeses would probably not have used speech as the language medium for training, said Rumbaugh, but we are fortunate, nevertheless, that they were able therapy researchers who tried to do what we now consider impossible, to teach a chimpanzee to use human speech.

The Gardners got away from these difficulties and helped to open up the possibility of two-way communication between man and ape. Not only did they find at the beginning that Washoe learned signs quickly; they also observed that she tended to change signs into a series, suggesting primitive phrases and sentences. This development suggested that apes might master syntax—a point, the speaker said he believes, that started up the other ape language projects. Briefly Rumbaugh described David Premack's work with Sarah, adding, "His basic strategy was and has been that every complex linguistic rule can be analyzed into simple units. The definition of those units and the teaching of them through appropriate methods can produce language confidence even in life forms that have no formal public language as employed by man." What his working group have with their own Lana project, said Rumbaugh, allows them to extend the basis for arguing that language is not a unique and distinguishing characteristic of man. Rather, it is "a competence which has its roots in the cognitive functions which men and apes share because of a close biological relationship." He summarized the conclusions he has come to:

First, that apes are relatively facile in the learning of words. They start out slowly but quickly become good at it, until frequently one lesson per word is enough.

Second, that apes readily start stringing signs or words together to form sentences. They can master the elements of syntax.

Third, that apes bring a readiness to extend their language skills—including syntax, or language structure—beyond the specifics of the contact within which their original learning occurs. If this were not true, probably the ape projects would be merely uninteresting demonstrations of rote learning.

Fourth, that apes can coin labels, and do so in a way that reflects their apparent sensibility to salient characteristics of the objects they label. For instance, as reported by Fouts, Washoe labeled a brazil nut "rock berry." Lana affords other examples: she has termed a Fanta orange soft drink as "the Coke which is orange," referring, of course, to its orange color; an overly ripe banana as "the banana which is black," and the (fruit) orange as "the apple which is orange" (again, in color).

These, said Dr. Rumbaugh, are relatively advanced coding processes strongly supported by these ape generative labels; they serve as important evidence that the ape's cognitive processes entail covert psychological functions of a linguistic nature. The central question is, are the apes' productions linguistic? There is no question whatsoever that they are communications that have an impact upon other animates within the commerce of social contacts, nor is there any doubt that they are adaptive in view of their problem-saving effectiveness. But are they tantamount to language use?

Of course, without a generally accepted definition of language the question cannot be answered definitively, but with or without the definition, many people are satisfied that the apes' productions are certainly relative to the language behavior of man. "Too many of the chimpanzees' productions have been modeled and appropriate to the context of the problem to be solved," said Dr. Rumbaugh. "Too many of them have been defined within their own context, unrelated to the specifics of events that immediately preceded their production. But how much is needed to say that the behavior of apes is qualitatively the same as the language behavior of man? To ask more and more of the apes and thereby to continue to escalate the criteria for the apes' productions to be called language is without constructive purpose."

It is certainly true, I reflected, that every time a new example is produced of the apes' ability, a chorus of dissent is heard, with angry scholars hiking the definition of language up, and up, and up.

In the meantime Dr. Rumbaugh thought it safe to draw the following conclusions:

First, that language as a form of communication is not totally unique from animal communication. Even with a formal public language, man uses many nonverbal forms of communication that do not differ substantially from those used by nonhuman primates, which is good evidence of the common evolutionary communication pyramid.

Second, that man's formal public language is based on processes that have emerged during the course of evolution. The apes, at least, have certain well-advanced psychological mechanisms that are immediate requisites to language, for if they are given appropriate training they show a remarkable propensity to develop those skills and extend them spontaneously, with impressive adaptivity.

Third, that within the animal kingdom there seems to be a continuum along which potential for formal public language becomes increasingly refined and powerful. But neither the public product nor the requisites to language are uniquely human.

The ape language projects discussed at this symposium, said Dr. Rumbaugh, have grown out of the methods Yerkes brought to his early work, the evolutionary perspective of comparative psychology and experimental psychology. How shall they affect our views? Clearly the projects serve to narrow the gap between man and ape. Though total silence still holds between the two species, the linguistic exchanges now happening will serve to underscore the close biological relationship between the two. They can neither demean nor diminish man, nor do they serve to elevate or demote the ape. Man and ape are both distinctive and unique. They shall remain thus, but perhaps they will gain a better understanding as to who one another is in relation to the other.

On a playful note he concluded: The possibility should not be ruled out that someday language-trained apes will accompany men to the field, to facilitate field research and the interpretation of the apes' signal systems. If so, there will be problems of credits and coauthoring of reports by man and his ape colleagues.

Peter Marler of Rockefeller University admitted that his talk was a bit hindered at the beginning because he, too, had intended to quote that passage from Yerkes. Never mind, he said: he would read it again, and he did so, to emphasize the bit, "We may not safely assume that they have nothing but feelings to express," et cetera. Marler announced that he is particularly interested in the relationship between sound production and emotional expression. There is a general idea, he said, that animal expression tends to be connected with what psychologists call the affective (i.e., emotional) processes, contrasted with the more symbolic processes typical of our own species. These alternatives are usually presented as being different in kind—but are they? The fact is often overlooked that speech is nearly always accompanied by affective components, as for instance in his speech at that very moment, with hand gestures, eyebrow movements, and contrasting tones of voice. To illustrate, he raised his eyebrows very high.

It bothered him as an ethologist, he resumed, that we so readily accept this affective component when it comes to animals and don't allow ourselves, when thinking of animals, to realize that they too use something like a meld of symbolic and affective signaling, even in nature. He enlarged on this thesis, showing slides of various primate species.

Emotional components, he commented, do not necessarily exclude symbolic functions. "What we have in animals is the same thing we have in human speech, a kind of melding of the two," he said. "The content of an animal's utterance is much richer than we often suppose."

Dr. Sue Savage is a very pretty primatologist I met originally at the chimpanzee colony in Oklahoma. For the past year and a half she has been working with three specimens of the very rare and interesting pygmy (or Bonobo or *Pan paniscus*) chimpanzee at Yerkes Center. (Dr. Yerkes's favorite animal, the genius Prince Chim, was a pygmy, though nobody knew about the species difference in those days.) Her talk was about natural gestural communication among these animals.

"Why is there such discrepancy between what a chimp can do with a little training and what he has done when left to his own devices?" she asked. There are at least two answers to this question, she said: First, there exists of necessity a partial misunderstanding of what it is exactly like to communicate with an ape on the part of those who haven't had the opportunity to partake in the experiments, or at least to watch. Also, much of what is said with a linguistically trained chimpanzee is interpretable only when the context is known in detail and when it is used as part and parcel of the linguistic interchange. There is such a wide discrepancy between our experience and that of a chimp that, unless the person is accustomed to communication with a chimpanzee, misinterpretations are bound to occur. Before the work with Washoe, people assumed that chimp-to-chimp communication was completely affective. Since then, closer observation has been made of their social structure in the wild, and it has been found that there are indeed prelinguistic communications made between the animals, though just how is still not known.

Dr. Savage decided to try to find out, and that's what she has been doing, interacting with the chimpanzees. Is the method vocal? Is it facial? These expressions with a chimp are no easier to describe than they are in humans. She has been studying spontaneous or untutored communication in her subjects. Because there seems to be more communication in sexual behavior than in any other, that is what she has concentrated on, and she has discovered that they were communicating with a large variety of signals about a lot of different things. A male soliciting sexual intercourse with a female, for example, might put his arm out in a wide from-the-back gesture, showing her that he wants her to turn so that he can approach her from the back, much as we might make a circular gesture when we want to teach a dance step to someone.

Roger Fouts, coming next, took as his subject Ameslan in Pan. It was obvious from what we had already heard, he said, that it is very important to establish a relationship with the chimpanzee you're working with. Yerkes himself always showed tremendous respect for his chimps. To attain success in this work, you must have a sound and thorough knowledge of the organism's biological and behavioral background, but unfortunately these points have long been ignored in psychology. Skinner used to paint his boxes black and ignored the relationship: with him it was merely input and output, ignoring the most important thing in animal behavior. For example, if a relationship is not established, the organism (i.e., chimpanzee) will be unable to breed. If it doesn't breed, progeny of course will not exist; if the species takes this tack, it can't survive. Relationship is the most basic form of animal behavior we can study. In the wild, as Jane Goodall has pointed out, the infancy of the chimpanzee is a time of social learning, as is the adolescence of the animal, when the acquisition of social relationships takes place. If you're being taught language you've got to have another individual at least to talk to, otherwise you won't have language.

"Because of the lack of attention to relationships," said Dr. Fouts, "I think experimental psychology historically has taken a rather myopic view of behavior, and perhaps it's time to throw the bath water out and keep the baby. The bath water would be the myopic perceptions of animal behavior in terms of structure. It has, I think, retarded the development of looking at complex social behavior and cognitive behavior in the chimpanzee and other organisms. You should take into account the gregarious nature of the chimp: they thrive on relationships."

He spoke approvingly of the methods used by Keith and Cathy Hayes, who raised Viki in their home as a baby, and of the Gardners, who brought up Washoe in the same way.

"Behaviorism likewise has emphasized learning rather than the biological makeup," said Roger. "Also it tends at times to ignore the fact that you have a developing organism on your hands. It's changing, day by day." Roger should know, I reflected. He has been with Washoe from the beginning, moving with her to Oklahoma from Reno, and accompanying her as she became an adolescent and then an adult. She has now become a mother, though the baby lived for only a few hours. Better luck next time, one hopes: other considerations apart, it is of the liveliest interest to all these observers to discover if an educated chimpanzee like Washoe will teach her infant the acquired Ameslan in which she is so proficient.

He discussed what one often hears, that when a chimpanzee has reached the age of six, it is dangerous to work with it any more because the body becomes fantastically strong and the temper uncertain. That is nonsense, said Fouts. "I think it's quite true that you can't work with them if you treat them like cows or rats," he declared, "but if you show the respect that's due to a higher primate, a very sophisticated organism in terms of social behavior, you certainly can carry it off. These negativistic statements are encouraging the myopic tradition in experimental psychology, by reducing the potential for exploring cognitive behavior. In other words, it tends to implicitly infer that we're limited to studying the language behavior in chimpanzees that are six years old or under. It's also bad public relations for what I think is the most exciting species that a comparative psychologist has to study. If people in Washington hear that you can't work with a chimpanzee more than six years of age, certainly they won't be willing to fund research that involves older chimpanzees. I think that these damaging statements result from, perhaps, ignorance on the experimenter's part, and also a lack of knowledge of the biological and behavioral development of the chimpanzee. Poor social environment, lack of stimulation, and lack of care and understanding for the chimpanzee, also poor rearing conditions, can result in this type of behavior.

"It's much the same as with children. When they're very young you are able to intimidate them because you're bigger than they are: you can tell them to sit up straight, mind their table manners, and so on. This goes well for a time. However, you've got to remember that they grow up and change, mature, and some people don't seem to be as aware of this as they should. As a result, when the chimp gets to be five or six years of age he begins to go through a rebellious period; the same thing, I think, happens with humans at a little later age. Certainly, if you began to treat a fifteen-year-old human like a five-year-old, you'd have a problem on your hands. Often when you have this with the human or chimpanzee, it results in a permanent caging for the chimpanzee and semi-permanent caging for the human. This is not to say chimps aren't difficult to work with; they're very interesting to work with and they're also very strong, and they do test the rules. No two days are the same. It's not like going down into the basement of the psychology building and dropping a white rat into the end of a maze.

"I've noticed that the teen-age phase in the human and the aging phase in the chimpanzee—we might call it adolescence in both cases—change as they mature. Your behavior has to change too. Instead of telling the chimp or the child, in parental tones, to sit up straight, it's better to sit down and discuss things with them rather than imposing your personal prejudices on them. The strength of the chimp is a problem, but there are ways around that. If I get into an argument with Billy or Washoe or some of the other chimps, I try to change the subject. I might do this by giving an alarm, pretending that there's a leopard concealed in the bush: the chimp might rush to protect me, thus preventing a confrontation."

He told of an experience he had when he had just arrived in Oklahoma to work with Dr. William Lemmon at the chimpanzee colony in Norman. In the laboratory was an approximately two-hundred-pound male named Lucifer. Roger, as part of his training, was cleaning out the laboratory, and he began hosing out Lucifer's cage. Lucifer got hold of the hose and started to pull from his end, dragging it into the cage. The hose was very expensive, and Roger was sure he would lose his job if Dr. Lemmon found that it had been eaten, so he tried hard to pull it away, but in vain. He knew he was losing. They kept a BB gun in the laboratory, so he grabbed that, aimed it at the chimp, and fired. Usually such a shot merely stings the animal but has no worse effect. That time, the BB shot rolled out of the barrel and fell to the ground. Then Roger looked past Lucifer and shouted, "Look behind you!"

Lucifer did, and Roger grabbed the hose and got it out of the way.

Another time he ran into trouble with Booie, who was off the island where he usually lived, serving to give practical training to a group of students. Roger was the experienced hand among them, and they treated him with great respect. When it was time for Booie to go back to the island, however, the chimp didn't obey orders but climbed a tree instead. Roger ordered him repeatedly to come down, but he wouldn't. He was on a long lead, so Roger wrapped the end of it around his hand and tugged. Booie promptly reached down with one hand, holding to the tree with the other, and lifted his lead, with Roger tied to it, off the ground, so he was swinging in the air.

"At that point I did what any comparative psychologist would do in such a situation," said Roger. "I forgave him for being a bad boy, whereupon he came down out of the tree and leapt into my arms. These things *can* be gotten round."

He mentioned several chimps that have been successfully raised to maturity, and the people who can handle them, Sue Savage being one. The celebrated Lucy is still living with her human parents, Dr. and Mrs. Temerlin, at eleven and a half years of age. "The father-daughter relationship has changed, though," admitted Roger. "He used to make her a whiskey sour in the evening, and now she's decided she likes his martinis instead, so he goes along with that."

He himself still works with Washoe, who is also eleven and a half. It has been said of the project that apes string their words together at random. Roger said he doesn't blame people for this: he was in France a year ago and he doesn't speak French very well, and he had the impression that the French randomly string their words together, too. "As far as I know," he said, "the only thing we have that approximates randomness is computers, in terms of generating random statements." Washoe had a definite word order in mind, and it wasn't ours.

Tim V. Gill told us about Lana and his conversations with the chimp. The computer routine had been followed for some time when he began to wonder what would happen if he introduced into the exchanges a number of misrepresentations of truth and similar problems. How would she deal with this unexpected turn? She was four and a half years old at the time and had been working with the computer for two years. Most of their communication had to do with food and drink. Lana was accustomed to getting milk in the morning—"Lana want what drink?" was the customary question, tapped out on the console—and a prepared concoction called "monkey chow" to eat in the afternoon. Now, as part of the experiment, sometimes she was given water instead of milk, and cabbage (which she did not care for) instead of chow. Whenever Tim varied the routine in this way, she noticed and protested. Sometimes she answered with her own variation; sometimes she simply refused, as when, in the afternoon when she would normally be expecting food, he asked, "Lana want what drink?" and she replied, "No Lana want drink."

In all instances when water was substituted for whatever drink she had asked for, she requested that the water be removed from the dispenser: for example, if water was given her instead of monkey chow, she asked that it be taken away, and when the experimenter read her question, "Water out of machine?" he said, "Yes," and removed it.

"You put chow in machine?" asked Lana.

"Yes," said Tim, and did so. End of conversation.

On every occasion when a substitution was made for what she had asked for, she acknowledged that fact. An interesting point was that whenever things came along normally in the way she was used to, she addressed the machine instead of Tim, but when something wasn't normal—as, for instance, when he put only half the expected monkey chow into the dispenser—she addressed him directly, as, "You put more chow in machine."

"Perhaps her success with these two requests that an unwanted object be removed gave her a feeling of great power," said Tim. "Up until the time of this experiment Lana had never been played with in this manner. Nobody had ever lied to her nor played tricks with her within the confines of the language system. Consequently, the potential of her language system to counteract the undesirable acts of others may not have been known to her, or may have been needed in order for her to discover it. In any event, she eventually learned that it was possible not to agree with the experiment, and, more important, to control it."

Clearly this is an example of what can be done by not treating Lana like a creature in a Skinnerian black box. She also seems to have grasped along the way the rudiments of tense, and how to use the negative. In addition, when Tim gave her water instead of the milk she asked for, after she tasted it she said, "Water name this."

"She made many novel and productive statements," said Gill in conclusion, "probably as a direct outgrowth of the experiments themselves. In these experiments Lana clearly demonstrated that she is operating in a domain once held explicit to man."

Two other experimenters with Lana read papers: one wanted to know what the chimp knew about numbers, and worked out a method by which she could tell, and the other found out that Lana could name objects, ask for their names when she didn't know them, and distinguish between their colors. She did very well on these tests.

Dr. Emil Menzel has worked for some years with chimpanzees and their methods of communicating with one another. "The problem I'm going to talk about today is an ancient and complicated one," he said, "but I think it can be stated in a straightforward fashion. Aristotle held that a falling stone accelerates as it nears the earth because it's becoming more jubilant at the approach of its natural home—just as a horse accelerates as it comes near the end of the journey. No scientist today finds it necessary to take the feelings or internal state of a falling body seriously. Why should we do so with a horse, chimpanzee, or man, for that matter? Well, some of us do." He reminded us of the experiences of Dr. Kellogg and his wife, who brought up Gua, a baby chimpanzee, with their child, and went on: "The social and philosophical attitude of the time in which we live is very important. Attitudes do change. A hundred years ago we were quite racist about other people, not to mention chimpanzees. We questioned whether some groups of humans were really people at all, though their languages were beautiful and complicated. Why has our attitude changed in this regard?

"This brings us to the third type of answer, which is empirical data," said Dr. Menzel. "Certainly empirical data is important, but I don't think it's the whole answer. All the evidence we have today regarding the humanness of other human groups was probably available one hundred years ago, but perhaps we just weren't prepared to see it."

Dr. Gordon Hewes explained why he, as an anthropologist, was attending this meeting of primatologists. It all began, he said, when as a young man in 1924 he persuaded the local public librarian in Santa Monica, California, to buy what was then a very expensive book, Yerkes's *The Great Apes,* which he had written with his wife: young Hewes was the first reader of it in Santa Monica. Indeed, after he got his degree in anthropology he had a few wild thoughts of going on and getting another in psychology, because of Dr. Yerkes and his work. He intended, he said, to talk about what all this meant to anthropologists, and then he embarked on a very erudite but amusing address. Since the middle of the nineteenth century, he said, anthropology has been unable to confine itself to the study of *Homo sapiens*, and after Yerkes started it was impossible to ignore him. It took a long time for anthropologists, however, to take an interest in apes except for their fossil teeth and skeletons.

The idea that apes might somehow learn to master human language goes back a long time. Descartes, of course, denied that animals could ever have any kind of language; that was in 1637. However, not long after that, in 1661, Samuel Pepys, the diarist, in England, having observed a large primate (which he referred to as a "great baboon"—as it could very well have been), thought it possible that such a creature "might learn to communicate by signs." In 1699 an anatomist, Herbert Tyson, who carefully dissected a young chimpanzee which had been brought to England and promptly died, expressed a great puzzlement over why such an animal could not speak, particularly as he had examined its vocal tract and brain and found them to be very manlike—the anatomy of a Pygmy, as he called it. In 1748 the philosopher de Lamettrie remarked that he thought apes might be capable of learning language. If he himself were to undertake such a project, said de Lamettrie, he would choose a young ape, one with an intelligent face, whose mastery of tasks had proved that he was intelligent: a good teacher for such an ape would be a certain man then famous for having taught deaf people, whose work had much impressed the writer. De Lamettrie thought that apes could be taught, literally, to speak, and if they were so taught they would know a language. He speculated on what man had been like before the invention of words and the learning of a language, and concluded that he was just another animal species, though one with less instinct. He viewed language skills as the origin of law, science, and the fine arts, and held that they served to polish the rough diamond of the human mind. Should the ape master language, speculated de Lamettrie, he would no longer be a wild man, a defective man, but a perfect man, a little gentleman. (De Lamettrie very likely had seen a live two-year-old chimpanzee owned by the French naturalist Buffon, keeper of the Jardin du Roi, which was later the Jardin des Plantes. The little ape survived for a while in Paris around 1740.)

In 1773 Lord Monboddo, a Scottish high court judge who was also a writer on language origins, supposed that apes could speak but perhaps had forgotten how, as they were very like to man biologically, and the question of the origin and progress of language was again raised. In 1779 the Dutch anatomist and artist Pieter Kuyper satisfied himself by dissection of an orang that such an animal was structurally incapable of human speech, and this approach was replicated not too many years afterwards by the very famous comparative anatomist Baron Cuvier, who had an opportunity actually to study a live orang. During the eighteenth and early nineteenth centuries generally, whenever chimpanzees and orangs were brought to Europe and, as usual, survived only a very short time, the people who observed and wrote about them almost always speculated about their possible language capacity.

Hewes talked at length about the Englishman, George Romanes, who wrote about animal communication, and also said something of the American Richard Lynch Garner, who, he said, "was almost pathetic in his conviction that he was decoding the speech of his primate friends," and who went to French West Africa in the 1890s, bringing with him the latest model Edison cylinder phonograph recording instrument. Garner hired a large cage, in which he sat in the tropical rain forest about two hundred yards from the place where he was staying, so that he might observe the apes and monkeys that visited his cage and communicated with him. Though his work was full of anthropomorphism and heavy sentimentalizing about apes and monkeys, Robert Yerkes recognized some important leads in his book and recommended that students read it.

Hewes continued. In Yerkes's textbook, published in 1911 and entitled *Introduction to Psychology,* he remarked, "Doubtless the psychological significance of the way in which language develops is of great importance," and that is all he said on the subject. Hewes added that it was strange that Yerkes didn't mention the work of Wilhelm Mundt, who in the late 1890s turned increasingly to the subject of language and culture. Mundt wrote of gesture language, sign language: nevertheless Yerkes didn't mention him. He did, however, correspond a good deal with Wolfgang Köhler. At the same time Dr. William H. Furness III, in Philadelphia, was making his celebrated effort to teach a young orang to speak, with the results of "a few unvoiced and poorly pronounced nouns." (Dr. Furness's orang learned to say "Mama," "Papa," and "up.") Then in 1915 came the first birth of a chimpanzee in captivity, at Quinta Palatino near Havana, in a colony maintained by Rosalia Abreu, who loved monkeys and apes and collected them. About nine or ten years later Yerkes was able to go and visit Madame Abreu's colony, then the largest privately operated colony of its kind in the world, although it was not scientific or experimental in any way.

"While all this was going on," said Dr. Hewes, "what was the response of anthropology? We have also a kind of vested interest in apes, at least traditionally: we compare skulls of apes, and teeth of apes and monkeys as well as of man as they become available. Not very much response, as a matter of fact. There were very few anthropologists beyond the physical anthropologists, whose business it was to look for similarities and dissimilarities in the bones and skeletons and occasionally some of the soft parts of primates, but rarely did they pay any attention to live subjects."

He noted Yerkes's attempt, soon given up, to replicate Furness's modest speech training, in this case with his two young chimps Chim and Panzee, about 1923, in New Hampshire—and here, he admitted, he had run into that quotation which had already been twice read to us. "But just around this time, this is what I want to get to, things *were* coming together after a long dearth of seemingly relative information—in fact, after a long period of a deliberate and ingenious hoax which had been foisted upon the anthropologists of the world, in the name of the Piltdown fraud," he said. "Evidence was finally coming to light which in the fullness of time, some twenty-five years later, would help to get some idea of the primate chronology straightened out, and get away from the idea that our ancestors were very comfortable, for a long period of time, going around with practically fully developed, modern-type brains, even though they may have had apelike snouts, which probably completely satisfied those people who, in their anti-Darwinism whether explicit or overt, wanted the greatest possible distance between us and the pongids. That was one of the functions, apparently, of the Piltdown hoax." In other words, to disconcert anthropology and science in general.

In his recapitulation of the coming together of primatology and anthropology, Dr. Hewes made a brief mention of Raymond Dart's work in South Africa, when Dart supplied the first description of one of the fossil forms, which, if not in *Homo sapiens'* direct line, at least seemed to cast a small question on the plausibility of the big-brained monster of Piltdown. Though there were a few anthropologists and anatomists who never accepted Piltdown man, some of the most distinguished people in the field were wholly taken in by it: these included the noted anthropologist Sir Arthur Keith.

Finally, in his outline of anthropology marching with primatology, Hewes spoke of Earnest Hooton, who carried the torch for anthropology even in the titles of his books, starting comparatively early. In 1931 he published *Up from the Ape;* in 1937 *Apes, Men and Morons;* in 1940 *Why Men Behave Like Apes and Vice Versa,* and then one where he finally dropped the term "ape" though it was still about apes and monkeys—*Man's Poor Relations,* in 1942.

An earlier speaker, Meredith Crawford, had spoken of an experiment at Orange Park in his day that involved cooperative problem solving by a pair of chimpanzees who had to haul a heavy fruit box toward their cage on separate ropes. Dr. Hewes now mentioned this again, reminding us that Crawford had said, "It may be that an important transitional step in the development of language behavior lies between the direct orientation of one animal by another through body manipulation and indirect confrontation through pointing." A film had been made of this experiment and we saw it at the end of the symposium. It was startling, even if nobody had spoken of it in advance. The moving spirit of the two chimpanzees kept exhorting the other by gesture to keep up the work, clapping him on the back, pointing, urging him excitedly, until at last the box came within reach. There could be no slightest doubt that those two chimps were communicating, and yet the film, so modern in its message, was made back in 1943.

Hewes wound up his talk with mention of one more anthropologist, A. L. Kroger of Berkeley, who had read Köhler and some of the publications of Nadia Ladygin-Kohts of Moscow: she had a female chimpanzee named Joni and wrote scientific papers about her. On the basis of this rather scanty reading, Kroger published a paper called "Subhuman Culture Beginnings" in 1928, in the *Quarterly Review of Biology.* Unfortunately he took the opposite view to Yerkes's on the subject of why apes don't talk. *He* thought it is because they don't have anything to talk about. Dr. Hewes recollected that Wittgenstein once made a comparable remark: "If lions could talk, we would not understand them anyway."

In conclusion Hewes remarked, "I am virtually certain that if Descartes had owned a young chimpanzee, he would have changed his mind."

Of course, after all the talk about Lana I was anxious to see her. Communicating chimpanzees are no novelty to me by this time, but my experience had always been with the ones that use Ameslan or a combination of Ameslan and old-fashioned comprehension of speech. Just how much spoken language a chimpanzee understands has never been thoroughly defined even by those who work with it, but I have seen some very surprising proofs of the process. Ally, an exceptionally intelligent young male of the colony at Norman, not only reacts to words spoken to his face but will (if he feels like it) obey instructions given to him over the telephone by Roger Fouts.

"Ally," Roger will say, "go fetch your bicycle and bring it over here to the phone," and Ally does just that.

There are other examples. One day Jane Temerlin, working in the Lemmon laboratory, lost an important key. She suspected that she might have dropped it in the big cage where a number of chimps were living, so she told Pan, the old man of the colony there, that if he found it and gave it to her she would reward him with a banana. Pan promptly went and picked up the key from an obscure corner, brought it over to Jane, and swapped it for his reward. I myself saw something similar at the Oklahoma City Zoo, not as it happens with a chimpanzee but with a full-grown female gorilla. I had been feeding her ice cream, and she grabbed the spoon and wouldn't give it back. The keeper who was with me went and fetched a fresh carton of ice cream, held it up where she could see it, and told her that he would exchange it for the spoon. She promptly handed back the spoon and got her ice cream.

Lana, however, was obviously a different proposition. There has never been any human-type conversation used on her.

Professor Hewes, his wife, and I went together to see her in the building at Yerkes Center where she has spent most of her life. The setup was just as has been described, with Lana, behind reinforced transparent walls, wandering back and forth near the console through which she communicates with the world. She had an impatient, busy air, her eyes fixed on Tim as he moved about the room outside. Dr. Rumbaugh, too, was present, and Lana seemed aware of him, but Tim was clearly the true interpreter for her. She duly tapped out requests for food and drink, and now and then, as if looking for recreation, she activated the machine into giving her picture slides that were thrown on a whitewashed bit of the wall. Now and then, too, she played music.

"Lana has her choice of various kinds of music, but she likes rock best," said Tim.

As the chimpanzee tapped out her demands, the teletype duly recorded whatever was said. Gordon Hewes was invited by Dr. Rumbaugh to ask the chimp questions, but her attention strayed because Tim and the visitors were drinking cups of coffee. Lana, too, asked for coffee, several times, but each time Tim said, "No." At last she gave more attention to Professor Hewes. Near the console but outside of the room was a table bearing several objects, which he was supposed to discuss. Dr. Hewes indicated a shoe on the middle of the table and asked, "What is that?"

"That is a shoe," tapped Lana. She went through the same

routine with a bowl ("The color of that bowl is purple," was a remark she offered as extra information) and with a box, but then for some reason she stalled. Hewes wanted her to tell him that a book on the table was red, but Lana resisted saying so. "The name of that is 'book,' " she replied, and again, when asked what color it was, "The name of that is 'book.' "

"She won't do any more now," Tim said. "She's bored."

Walking about, I studied a cubicle similar to Lana's but set farther back in the big room. It was occupied by four very young chimpanzees who tumbled about and paid little attention to what was going on outside.

"That's our next project," said Tim. "They're going to be trained something along the same lines as Lana. The program won't be identical, though. We keep thinking of improvements."

Dr. Rumbaugh added, "It's hoped that this type of training may have usefulness when it's applied to the treatment of retarded children. Time will tell."

Up at the far end of the room some more primatologists had entered, and my party joined them. I didn't, because I wanted to see more of the baby chimps in their nursery. After a bit I started back toward the teletype and paused for a last glimpse of Lana, who seemed to be amusing herself watching slides. If her eyes slewed to the side to take me in, and I think they must have, I didn't notice. All I saw was that Lana had begun some rather strange behavior, shifting from one foot to the other. It should have warned me, but my mind was still on the infant chimps.

Suddenly Lana's body hurtled through the air at me and landed against the wall with a ferocious crash. I was expected to jump away in fright, which would have pleased her. But my reactions are always lethargic, and I didn't move except to blink. Disappointed, the chimpanzee picked herself up, wandered back to the console, and idly punched out some rock.

Oh well, I reflected, no doubt she wanted a little excitement.

Bibliography

Blake, Henry. *Talking with Horses.* New York: E. P. Dutton, 1976.

Borgese, Elisabeth Mann. *The Language Barrier: Beasts and Man.* New York: Holt, Rinehart and Winston, 1965.

Christopher, Milbourne. *ESP, Seers and Psychics.* New York: Thomas Y. Crowell Co., 1970.

Fox, Michael. *Behavior of Wolves, Dogs and Related Canids.* London: Jonathan Cape, 1971.

Griffin, Donald R. *The Question of Animal Awareness: Evolutionary Continuity and Mental Experience.* New York: Rockefeller University Press, 1976.

Lenneberg, Eric H. *Biological Foundations of Language.* New York: John Wiley & Sons, 1967.

Lilly, John C. *Man and Dolphin.* New York: Doubleday, 1961.

Pfungst, Oskar. *Clever Hans.* New York: H. Holt & Co., 1911.

Romanes, George J. *Animal Intelligence.* New York: D. Appleton & Co., 1883.

————. *Mental Evolution in Animals.* London: K. Paul, Trench & Co., 1883.

Rumbaugh, Duane M., ed. *Language Learning by a Chimpanzee: The Lana Project.* New York: Academic Press, 1977.

Sebeok, Thomas A., ed. *Animal Communication: Techniques of Study and Results of Research.* Bloomington: Indiana University Press, 1968.

Singh, Joseph A., and Zingg, Robert M. *Wolf Children and Feral Man.* Hamden, Conn.: Shoe String Press, 1966.

Wilson, E. O. *Sociobiology.* Cambridge, Mass.: Harvard University Press, 1976.

Index

HOW TO WRITE
REPORTS

ANNE FAUNDEZ

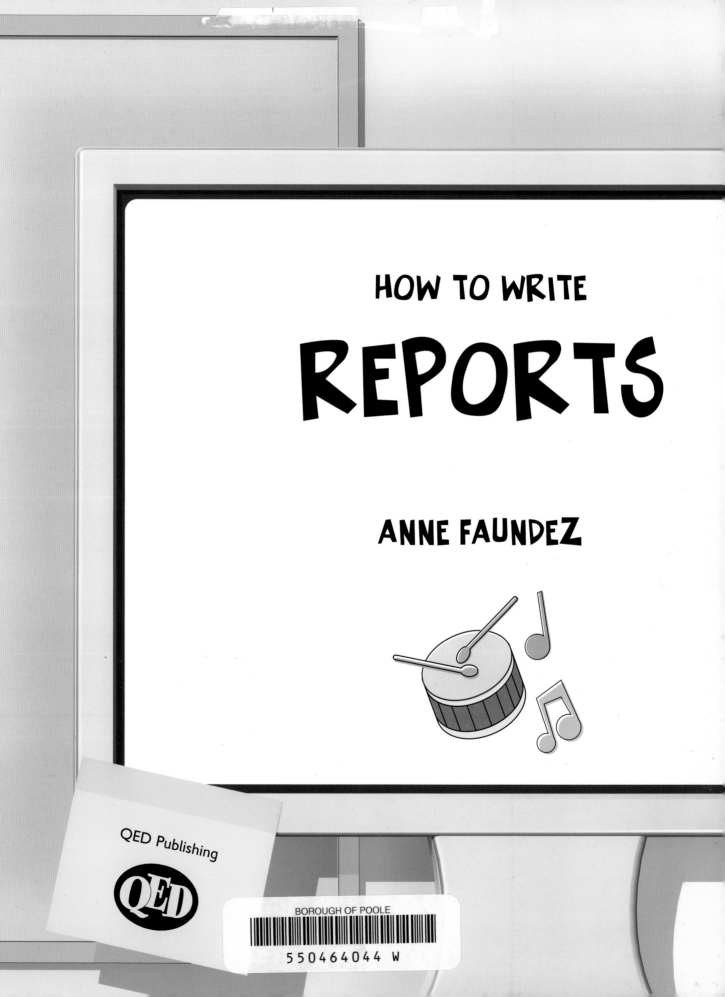

QED Publishing

QED

First published in the UK in 2007 by
QED Publishing
A Quarto Group company
226 City Road
London EC1V 2TT
www.qed-publishing.co.uk

A Catalogue record for this book is
available from the British Library.

ISBN 978 1 84538 906 2

Written by Anne Faundez
Designed by Jackie Palmer
Editor Louisa Somerville
Illustrations by Tim Loughead

Publisher Steve Evans
Creative Director Zeta Davies
Senior Editior Hannah Ray

Printed and bound in China

Words in **bold** are explained
in the glossary on page 30.

Website information is correct at
time of going to press. However, the
publishers cannot accept liability for
any information or links found on
third-party websites.

CONTENTS

NON-FICTION TEXTS

When you write about yourself and the world around you, giving information, facts and, sometimes, personal opinions, you are writing non-fiction. There are all sorts of non-fiction texts. A newspaper report, a **CV**, a **review**, a **recount** of a holiday, a balanced argument and a project on the rainforest are all examples of non-fiction writing.

Purpose

The words you use and how you present your writing depends on what sort of non-fiction you want to write. Are you writing for a class newspaper? Then you need a catchy **headline**. Are you doing a project about the Ancient Greeks? Then you need to organize the information under **subheadings**, with each subheading introducing a new aspect of the topic.

Audience

Who will be reading what you've written? If your writing is intended for your friends, then you can use language that they are familiar with and word **contractions**, such as 'doesn't' and 'won't'. If your writing is for your teacher or someone you don't know, then you should use formal language – 'will not', for example – and avoid words like 'brilliant' and 'cool'.

A class diary is an example of a non-fiction text.

glue stick

Class 6 Diary
My birthday treat
by Leonie Smith

June 10th
For my birthday, my dad took me to visit Zone Studios on Saturday. Galaxy Gorgons, my favourite film, was made here.
In the first building were the sets from some of the scenes. In one corner, we even saw the gorgons' spaceship! We entered a workshop where people were working on a model of a monstrous one-eyed fish. Maybe this character will star in the next Galaxy film!

Then we went into the props and costumes department – and saw the necklace with extra-sensory powers! Finally, we passed through a room full of computers where people were playing with on-screen graphics. This was my best ever birthday treat.

THEY CAME FROM OUTER SPACE...
the Galaxy Gorgons
STARRING

Researching

As non-fiction writing is about facts, you need to research your information. Talk to people with knowledge or experience, and use books or the Internet. Keep a notebook of interesting facts that you can use in later writing. Make sketches, too, if you like.

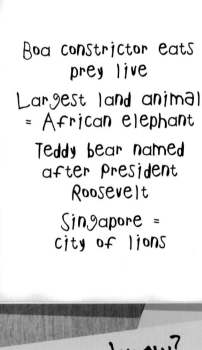

Boa constrictor eats prey live

Largest land animal = African elephant

Teddy bear named after President Roosevelt

Singapore = city of lions

Tip

Nobody can get it all right the first time they write. You need to plan and prepare an outline, either as a diagram or chart, or as a list, before writing.

Once you've written a **draft**, you may want to use a word-processing program to move **paragraphs** around or to do a **spell-check**.

Here are some good websites to use when you are doing your research:

www.thebritishmuseum.ac.uk

www.bbc.co.uk/cbbc/

www.enchantedlearning.com

Did you know?

Leonardo da Vinci filled his notebooks with sketches of all sorts of things: a bird's wing, devices to make water flow, hats. He also did mirror writing, going backwards across the page to form a mirror image of his normal writing.

WRITING ABOUT A TRIP

A piece of writing that retells events in the order in which they happened is called a recount. Suppose you've been on a school trip and your teacher has asked you to write about it. The best way to do this is to describe the events as you experienced them.

Starting to write

Present the events in **chronological order.** Describe briefly the place you were visiting so that the reader can get a flavour of it. Since you were the person who experienced everything, write from your point of view, either using the pronoun 'I' or, if you were with your class or friends, 'we'. Make your writing interesting by adding your own observations and feelings.

Which tense?

You should write in the past tense because you are describing events that have already taken place.

All in order

Before you start, it's helpful to draw a timeline. Write down the events of the trip in the order that they happened from start to finish. You only need to write key words or abbreviations that will help you to write your recount.

Useful words to use that show the sequence of events:

- First
- Next
- Then
- Shortly after
- Later
- Finally

Start with a sentence that sets the scene.

Each paragraph explains a different event in chronological order.

Try to include some unusual facts to keep your reader interested.

Notes for trip to Miramar
Guide – Tunnel – Reefs (coral) –
Coast (sharks!) Shipwreck – Dangerzone
(snakes) – Talk

Give your recount a title that tells the reader what it is about.

Break the text into paragraphs to make it easier to read.

Our trip to the aquatic life centre

Today we visited the Miramar Aquatic Life Centre. Our coach pulled up just as it was opening. Inside, our guide was waiting for us. We followed him down into a transparent tunnel, with the sea all around. There were real-life fish swimming right up to us.

First, we visited the Reefs Sanctuary, with hundreds of rainbow-coloured fish darting in and out of bits of red and pink coral. After that, we moved on to the Coast Sanctuary, which had huge stingrays, turtles, sharks and even a shipwreck. One shark came swimming right up to me. I had never been face to face with a shark before and I was so glad that I was safe in the tunnel, away from its huge teeth.

Next, we visited the Dangerzone. It had sea snakes and gigantic jellyfish.

After lunch, we listened to a talk on some of the fish we'd seen. It was really interesting. Did you know that the jellyfish has no brain?

By the time we'd finished we were tired and ready for the ride home, but it had been a great day!

After retelling the events, end with a closing **statement**

Springboard

Write a recount of a family outing or a trip to a museum. Begin by making a timeline of the main events. Don't forget to use some time-sequencing words (first, next and so on).

NEWSPAPER REPORT

Newspaper reports have to grab the reader's attention. They need punchy headlines and a gripping first few lines to make sure that you read on. News stories cover everything from world events to lost puppies. In the same way, you can turn anything exciting or unusual that happens to you into a news report.

News style

Flip through some newspaper **articles** and you'll see that they have things in common:

The headline (title) is set in large type. It consists of a few words only — just enough to catch the reader's eye.

the daily news

WALKING SHARK

Scientists discover a shark in the coral reefs off Indonesia that can walk on its fins

By Shah Kattak

A team of scientists has discovered many new species of fish and coral around the islands of Indonesia – including a fish that can walk.

Legging it
The shark is just over a metre long, with a slender body. It uses its pectoral fins as 'legs' to walk along the sea floor in search of food. Sam Sebastian, a leading member of the team, said, "They're extraordinary animals that sort of walk on their fins. They spend a lot of time on the bottom looking for mussels and crabs."

Watch the birdie
The coastal area, known as Bird's Head, is home to more than 1200 species of fish and almost 600 types of coral.

The introduction is one or two sentences long and says what the article is about.

The name of the person who wrote the article is known as the byline.

The paragraphs are arranged under subheadings.

The text is broken into short paragraphs.

8

What to write?

Try writing a news report about something that has happened to you recently – the day your mum went to work in her slippers or when your school won a prize, for example. To catch the reader's eye you need an attention-grabbing headline. Make sure that it describes what your article is about. Use rhyme, **alliteration** or even puns – and keep it short!

Tackling the facts

A newspaper report describes what happened, who was involved, and when, how and where the event took place. Introduce the subject in the first paragraph. In the next paragraph build on what you have said, adding supporting facts, background information and **quotes**. Put less important details at the end of your article. For example, you might start with:

Headline uses alliteration.

JONES' JOURNEY JEOPARDY
Ten-year-old Aaron Jones had a brush with danger on his way to school.

First line makes you want to read on.

Tip

Use a computer to lay out your draft article in the style of a newspaper **column**. Vary the **fonts**, setting the headline in the largest font size and in **bold**. You'll need to **edit** your article to make it fit the column.

Writing style

Make your article exiting by varying your writing style.

- Follow a short sentence with a longer one.
- Use active verbs ('a storm lashed the town'), not passive ones ('the town was lashed by a storm').
- Introduce different points of view by using quotes from the people involved.
- Divide your writing into paragraphs of three or four sentences.
- Use subheadings.
- Break up your story with illustrations – either photos or drawings – and informative **captions**.

Did you know?

Newspapers have been around for a very long time. In Ancient Rome, the government kept its citizens informed of the latest developments in a war or the emperor's health through hand-written newssheets. These were placed in market squares where everyone could go and read them.

Springboard

Rewrite a nursery rhyme or fairy story as a newspaper report. Here's a headline to get you started: 'Eggs-hausting', says Humpty Dumpty.

AN INTERVIEW

An interview is a meeting with someone to ask them questions, either about themselves or on a special topic. The interviewer is the person who asks the questions and the interviewee is the person who answers them. The interviewer takes notes and then writes them up as a non-fiction text.

Who to interview?

You might wish to ask a local author about their writing, question your grandparents about life 50 years ago or even find out what your neighbours think about the shopping mall that's just opened nearby.

Preparing an interview

Make a list of questions. If you are interviewing people about their past life, your questions might be about the clothes they wore then, the music they enjoyed, their hobbies, or even how they got to school in the morning. Make sure you phrase your questions so that the person is encouraged to speak, and not to give 'yes' or 'no' answers.

Interview with Daz Dayman

Q. How did you feel when you got home?
A. v. happy & good 2 c my cats. Not v. much time 2 relax as next album 2 rcd.

Q. What do you think about a reunion?
A. wd like 2 c band m'bers agn

Q. What sort of music did you listen to?

A.

Q. What was it like...?

A.

Q. Tell me about...

A.

Abbreviate words as you write down answers.

Allow enough of a gap to write down your interviewee's answers.

ACTIVITY

Write down the questions on a notepad, leaving plenty of space between each one. Use this blank space to write down your interviewee's answers.

Interviewing

Now that you've got a list of questions, set about finding the right person to answer them. If your project is about life in the 1950s, the best people to interview are grandparents and other older people. If you want to find out what people think about the new shopping mall, ask people who live in the area. Try to write down as much as you can so that you will be able to convey the character of your interviewee.

Writing up the interview

Begin your interview with an introduction, saying who the interviewee is and what the interview is about. Set out your interview like a playscript, with a new line for each speaker. Make the interview sound as if the interviewee is talking by using the actual words he or she used. Don't forget to replace your abbreviations with full words!

Begin a new line for each speaker.

Use initials to show who is speaking.

Include expressions that the interviewee uses, such as 'um', 'er' and 'oh', to make it sound as though they're speaking.

BEATLES' FAN ELLA EVANS TALKS TO HER NIECE, KIM, ABOUT GROWING UP IN THE 1960S.

KE: What did you like and dislike about school?

EE: My favourite lesson was geography. I loved learning about other countries. In those days, we didn't go on holiday abroad, so even learning about France seemed exotic! My least favourite subject was sewing. It would take me the whole lesson just to thread a needle!

KE: What did you want to be when you were my age?

EE: When I saw on television someone walking on the moon for the first time, it so impressed me that I wanted to be an astronaut.

KE: What sort of music did you like?

EE: My favourite music was rock-'n'-roll – oh, and anything by The Beatles.

Springboard

Make a list of questions you'd like to ask one of the following people:

- your favourite pop star
- a sporting hero
- Christopher Columbus
- a character from a book

BOOK REVIEW

A review is a report in which you give your opinion about, for example, a book, film, CD, concert, play or an exhibition.

Write a book review

How you write will depend on whether you are reviewing a **fiction** or non-fiction book. However, your review should always start with the title of the book, the name of the author and, if the book has pictures, the name of the illustrator or photographer.

Review a story or picture book

Tell the reader where the story takes place and describe the main characters. Write about the characters' relationships to each other. Are they school friends or did they meet through some strange coincidence? Describe the **plot** – but not in too much detail. Now give your opinion of the book. What do you like about the story? Is the plot convincing? Are the characters believable? Is it a book that you couldn't put down? If the book has illustrations, what did you like about them? Did they catch the mood?

Say what the book is about.

Put the name of the book, the author and the illustrator at the top.

Mr Whizz's School of Magic

by Ebony Lee
Illustrated by Jo Melon

This story is about a group of pupils at Mr Whizz's school for wizards. Most lessons involve boring experiments such as how to turn a pea into a pumpkin. But one day Mr Whizz walks into the classroom with a brand-new book of spells. From that moment on, life changes dramatically, especially for star pupils Jazz and Glitter – and, of course, Mr Whizz.

My favourite characters are Mr Whizz, who is always forgetting things, and Glitter, who is warm-hearted, clever and funny.

The black-and-white drawings have lots of detail and show the characters just as I imagined them to be.

I would recommend this book to anyone who enjoys a good laugh.

Mention your favourite character.

Who are the main characters?

Say what you liked about the story.

What do you think of the illustrations?

Don't give away the ending. If your review has been successful, the reader will want to find out for themselves.

Review a non-fiction book

Reviewing a non-fiction book is a little different. Write a **summary** of the contents. Describe the way the book is organized. Does it have a **glossary** and an **index**? Does it have photos or diagrams to support the text? Now give your opinion of the book. Is it clearly written? Does it hold your interest and tell you what you wanted to know? Are the photographs eye-catching and informative? Do you know more about the topic after having read the book? Does the book make you want to read more about the topic?

Put the name of the book, the author and the photographer at the top.

Endangered Animals
By Joshua Squire
Photographs by Tammy Wayne

How are the contents organized?

Say what the book is about.

The book looks at animals, from every continent, that are in danger of extinction.

The book is divided into sections, one for each continent. Five types of animal – mammal, fish, bird, reptile and insect – are examined in each section. Under 'Australia' for example, you will find pages on the koala, saltwater crocodile, giant dragonfly, turquoise parrot and lungfish. The text is divided into subheadings that give facts about each animal and an explanation of why it is under threat.

Describe photos and charts.

The book has lots of large, really interesting photos. There's a map at the beginning of each section that shows where each animal comes from.

Mention any special features.

The book is well organized and clearly written, with an index so that you can find facts easily. It is so informative that it makes me want to find out more about the topic.

What did you like about the book?

13

BALANCED REPORT

A balanced report presents both sides of an argument and is usually made up of facts, or a mix of fact and opinion. Often the issue is **controversial**, with people either for or against it.

What's the purpose?

There are several reasons for writing a balanced report:

- To show that there are two sides to an argument.
- To give readers all the information they need to make up their minds about where they stand on an issue.
- To have all the facts in front of you before you take sides in a debate.

Facts, facts, facts

You'll need to research your topic thoroughly to find out the facts for and against an argument before setting out your report. Your research might include:
- Finding facts on the Internet or in your local library.
- Conducting a **survey** or **questionnaire**.
- Interviewing people.

Writing style

When you write up your report, use the present tense ('need' instead of 'needed') and use **impersonal language** ('it was found' instead of 'we found'...)

ACTIVITY

Try preparing a balanced report. There's a rumour flying around that a theme park is going to be built just outside your town. Some of your friends favour this proposal, others are against it. You're not sure where you stand so you need to gather all the facts. What arguments can you come up with, for and against?

Useful words

However • On the one hand/On the other hand • Therefore • Then • To sum up • To conclude

Springboard

Choose one of these topics and produce a list of arguments for and against it. Try to have the same number of points in each column.

- Should school uniform be compulsory?
- Is tourism good for a country?
- Should zoos exist?
- Is reality TV entertainment?

Tip

Use bullet points for each new argument. This will make the information easier to read.

CLASS 6 REPORT

SHOULD THERE BE MORE SPORT IN THE SCHOOL DAY?

Class 6 reports on the arguments for and against more sport being introduced during the school day.

Begin with an opening statement to tell the reader what the report is about.

Use impersonal language and the present tense.

REMEMBER TO:

List 'for' and 'against' arguments in separate columns.

Write a point in favour of the argument and follow it with a point against.

FOR

- Children need more exercise, especially now that they are driven to school and spend free time watching TV and playing video games.

- Being overweight is a growing problem.

- There's too much time spent on subjects such as maths, so that children who are good at other things, like sport, lose out.

AGAINST

- It's not up to the school to provide more sport. Parents need to get their children to be more active.

- It would be better to give more time to teaching children about healthy eating instead.

- To cover all the subjects, the school day would have to be lengthened.

End your report with a statement summing up both arguments or stating which argument you agree with.

Place the 'for' and 'against' arguments alongside each other so that they roughly match up.

To sum up, after reading the arguments, Class 6 believes that it would be a good idea for children to play more sports during the school day.

SPORTS REPORT

Suppose that you've been asked to write about a school sports event – a swimming gala, the football finals, or a sports day event, such as the 100-metre race or even the egg-and-spoon race. How can you make your writing grab the reader's attention?

Starting to write

Here are some tips for spot-on sports reporting:

- Keep sentences short and avoid **explanatory clauses**. For example, instead of writing 'All eyes were on Hari, who, as well as being really popular, was the school's best sprinter', you could say 'All eyes were on Hari. He was the school's best sprinter. He was also really popular'.

- Use an **active voice** ('Ben pounded the ground'). A **passive voice** ('the ground was pounded by Ben'), will slow down your writing.

- Use **metaphors** and **similes** – 'Anya was a white streak in the distance', 'Misha raced like the wind'.

- Build mood. For example, describe the suspense before the race. Which of the entrants looked nervous/confident? Were the spectators noisy with excitement or hushed with suspense? You could also describe key moments of tension – the moment when Jay's dog slipped its collar one minute into the race and rushed along the track, almost tripping her up.

Springboard

Write a report about your school team's most recent match. (Choose any team sport, such as football or netball.) Write the report as if you are writing for a local newspaper and the team members are local sports stars. Include quotes from the team and from their manager, of course.

ACTIVITY

Choose a sport to write about. How are you going to present it? Are you going to start with the result, 'Anna wins by a whisker'? Or do you wish to write it as the event takes place, keeping your audience in suspense until the very end? If so, describe the sequence of events as they happen.

Organize your writing into short paragraphs to make it easier to read and hold the reader's interest.

Use adjectives, but use them sparingly as too many will slow down the impact of your writing.

Use powerful verbs – for 'ran', say 'galloped' or 'sprinted'.

United win
by Joshua Squire

A brilliant strike by newcomer Fernando saves the day for reigning champions, United. With no goals scored, United seemed content with a draw. But with just five minutes to go, Fernando saw his chance and grabbed the ball.

In for the kill

He cruised his way down the field and skilfully fought off one opponent after the other. With a superb flick, he winged the ball home. "I never thought I'd do it", he said triumphantly.

Use a subheading to introduce a new idea.

Use quotes from competitors and spectators to back up your statements.

Dull... or exciting?

Here are two reports of the same event, but one is boring while the other grabs your attention. What makes the difference?

explanatory clauses

Record finish in freestyle finals
by V. Dull

The freestyle race began. The lead was taken by Emma, <u>who was in lane 1</u> and whose dive placed her in front. <u>She swam through the water</u> quickly.
Then, from lane 3, Verna came forwards. <u>She swam through the water,</u> moving her way forwards with <u>strong</u> arm strokes and feet movements.

boring adjectives

repetition

active voice

simile

adjectives

powerful verbs

Record finish in freestyle finals
by V. Exciting

Splash! They <u>were</u> off! Emma in lane 1 got off to a great start. Her long dive put her into the lead. She glided effortlessly through the water, leaving behind a froth of <u>bubbles</u> <u>like a jet stream.</u>
Then Verna streaked forwards from lane 3. Soon she was inching her way forward with <u>swift</u> arm strokes and feet movements that barely <u>ruffled</u> the water's surface.

RECOUNTING EVENTS IN DIFFERENT STYLES

Very often, what we write and how we write it depends on who we are writing for. For example, you would use different styles of writing to describe a visit to a theme park to your best friend or to your teacher.

Informal style

Your best friend Nadine loves going to the theme park. You'd write to her in a natural, chatty style – almost as if you were talking to her. You'd take for granted that she already knows lots of things about the theme park, so you wouldn't have to explain the rides to her.

FUN PARK
ADMIT ONE

FUN PARK
ADMIT ONE

REMEMBER TO USE:

25, Diamond Lane
St Albans
Herts
20 Nov 2006

Dear Nadine,
    I went to Adventure Towers yesterday with Mum, Dad and Susie. The coach ride took ages (as usual). The rides were fantastic as ever and there was a gigantic new roller-coaster that you're going to love. I thought it'd be scary but it was really awesome.
    For lunch we had pancakes, and that ice cream you adore. Yum! In the afternoon we went on a water ride. You should have seen Dad – he got soaked. Mum couldn't stop laughing.
    We had a really cool day. I'll ask Mum if we can go again when you're back from hols.
Love Jane x x x

chatty, informal style

colloquial language ('scary' and 'awesome')

snippets of information about the writer

exclamation marks for emphasis

contractions ('you're' instead of 'you are')

Formal style

Your teacher asks you to write about your outing for a display of work aimed at parents. As you don't know the people you're writing for, the language you use should be more formal. You need to give more details since they may not know the theme park. To help the reader, you could include some subheadings.

Give facts rather than personal opinions.

My trip to Adventure Towers
Yesterday, my parents took my sister and me to Adventure Towers, <u>the largest theme park in this part of the country</u>. It is not always easy to find parking there, so we went by coach.

All sorts of rides
We tried out lots of different rides. Then we went on the new roller-coaster. It was huge and had lots of carriages with four seats across each row. It was frightening but also great fun.

Water everywhere
We had lunch in a restaurant overlooking the lake. After lunch, we went on a water ride that took us up and down slides and through tunnels. We were thoroughly wet by the end of the ride.

Time for home
We were too tired to do much more after the water ride. Luckily we <u>did not</u> have to wait too long for the coach back into town. We all enjoyed ourselves and had a very exciting day.

Use formal language.

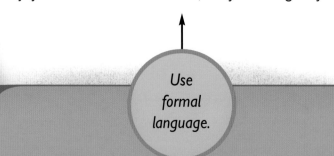

ACTIVITY

Write recounts of a birthday party you've enjoyed for two people from this list:

- grandparent
- younger brother or sister
- pen friend from abroad
- school magazine editor

Springboard

Keep a diary. Each day, recount the day's events. Decide if you'll write in an informal or formal style. If your diary is 'for your eyes only' you may choose an informal style. However, if you want others to read it eventually, you could be more formal.

WRITING A CV

A CV gives factual information about your education, achievements and hobbies. It will help someone decide whether you are the most suitable person for a position, for example to take part in a scheme or be given an award, or, when you're older, to be offered a job.

Did you know?

The letters CV stand for the Latin words curriculum vitae, which mean 'the story of your life'.

Is a CV the same as an autobiography?

A CV gives the facts about what a person has done in their life. The facts are usually set out in reverse order with the most recent first.

An autobiography is the story of someone's life, written by that person. It includes observations and feelings, facts and opinions. It is written in the first person: 'I'. The facts are usually set out in chronological order.

Writing a CV

It's important that you keep your CV short and to the point. Don't forget that the person reading it will also be reading lots of other CVs. Here are some tips:

• Begin by writing your name, date of birth and contact details.
• List all the things you've done, both in and out of school. Start with your most recent achievement.
• Include as many things as you can that are relevant to what you are applying for. You need to persuade the person that you are the most appropriate candidate.
• List your favourite hobbies and pastimes.

Do include in your CV achievements, such as:
• I was awarded my gold swimming badge.
• My painting was selected for a competition.
• I helped organize a 'bring and buy' sale.

DON'T include these in your CV!
• My cat loves cheese.
• My pencil case is blue.
• I failed my maths test.

Springboard

Put the title 'Curriculum Vitae' at the top. Centring the type on the page looks good.

Curriculum Vitae

Name: Toby Mann
Date of birth: 4 April 1998
Address: Flat A, 66 Wood Avenue, Goodtown XX12 5RY
Telephone: 00101 5252

List your educational achievements from the most recent, ending with the earliest.

Education
April 2007: Passed with honours Piano Playing and Composition, Grade 1
January 2007: Won third prize in Goodtown' s Music Festival, 12 Years and Under category
December 2006: Played Jack in school production of *Jack The Giant Killer*
May 2005: took part in interview about my school, published in the *Goodtown Gazette*

Hobbies:
Skateboarding, music and reading.

Include the names of two people who can recommend you, when you apply for a job.

References:
Mrs P. Shooter, Mr C. Saw

Tip

Type your CV on a computer. It looks neater and it can be easily updated and altered.

POWER OF PERSUASION

Writing that tries to persuade the reader to agree with what is being said is known as exaggerated writing. It is **biased** – it puts forward only one point of view. For example, a brochure advertising a holiday in a mountain chalet 'with magnificent views to the lake' avoids saying that you can only see the lake if you stand on a chair. An advert for an MP3 player claims that you can tune in to pop bands' live performances, but doesn't state that first you must buy an expensive card to slot into the player.

Twinkletoes
does the dancing for you!

Want to become a brilliant dancer in just TWO minutes? Then strap some Twinkletoes onto your shoes and get tapping.

You'll be a whirling, twirling disco star - and the envy of your friends.

Twinkletoes strap-on lights offer you the latest innovations in dance and rhythm technology.

Full money-back guarantee if you're not 100% thrilled with the results!

Order now while stocks last!

When to use exaggerated writing

People use exaggerated writing to try to convince the reader to, say, buy something that they are selling, such as a holiday, a new type of drink or a computer game.

Presenting exaggerated writing

Have a go at some exaggerated writing. Set out your writing in the form of an advertisement, a poster or a brochure, and make it as eye-catching as you can to grab the reader's attention.

• Use large and striking pictures.
• Vary the font size of your text.
• Write in the present tense to make it sound immediate and urgent.
• Use **positive** language that makes the reader think their life will be improved.

Word power

Choose your words carefully. Each word must make an impact. Rather than saying that something is 'good', use words such as 'brilliant', 'amazing', 'unique', 'wonderful', 'superb', 'fantastic' and 'one in a million'.

- Use lots of adjectives and superlatives – words such as 'idyllic', 'outstanding', 'coolest'.
- Present opinions as facts: 'Mrs Goochi says, "This is the best handbag ever!"'.
- Give your writing the stamp of approval of so-called experts with phrases such as 'leading experts agree', 'leader in the field recommends' and 'endorsed by'.
- Use technical or scientific words that sound authoritative, even though the reader won't understand them.

Use a software program to make a poster that looks really slick and professional. Set your text in different colours and fonts and vary the font size. You could advertise one of the following or invent your own product.

- A holiday destination
- A 'unique' hair product
- An environmentally-friendly car

Springboard

Put the most important information at the top.

Provide a quote to back up your statement.

Give details of where to purchase tickets at the bottom of the poster.

JUMPIN' BEENS

Live in concert
Your chance in a million to see the stars perform!

ONE NIGHT ONLY
8 May

The Disco Centre is proud to present the sensational Jumpin'Beens performing hits from their latest album, Tweenie Beenies

'Tweenie Beenies, with its unique mix of up-beat backing tracks and amazing lyrics, is a winner...the best album to hit the music scene in years' Music Review

To book your ticket visit www.jumpin'beens.com or call The Disco Centre on 145 39 09

INFORMATION WRITING

Imagine you've been asked to write a report on 'Animals of the Rainforest'. It's a topic that interests you but you don't know much about. Where do you start? Do you want your report to be about different types of mammals or all sorts of different animals including insects, reptiles and birds?

ACTIVITY

Have a mindmapping session with a group of friends. Write the words 'Rainforest animals' in the middle of a large sheet of paper. Then write on the names of the rainforest animals and what you know about each one. When your spidergram is finished, on another sheet of paper list the things that you need to research, such as: what other mammals live in the rainforest?, are there any rainforest insects?

Researching

Find out about your chosen animals, using information books and the Internet. If you're using books as reference, use the contents page and index to speed up your research. **Scan** the text for key words: 'Poison-arrow frogs are brightly coloured. Their colouring is a warning to other animals that they are poisonous.' Take notes. Write down only the words that are important e.g. frilly lizard = green, eats mosquitoes, hibernates.

What we know
Monkeys, toucans, frogs live in rainforest

What we need to find out
What is a rainforest/where is it found? (for intro)
Insects
Reptiles
Other mammals

Starting to write

Organize your writing into sections so that, for example, all your information on toucans is under the same section.
• Use a subheading for each section.
• Write in the present tense since you are describing the way the animals are.
• Move from the general to the particular, for example 'Monkeys communicate with each other by making loud noises. Howler monkeys are the loudest monkeys of all.'

Useful words

Use words that relate to your chosen topic to make your writing sound accurate and factual. For example, for a project on rainforest animals the following words would be useful: canopy, camouflage, extinction, leaf litter, species.

Use technical words, such as 'habitat', 'herbivores' and 'primates'.

Start with general information.

Tip
You don't have to arrange the sections of your report in any particular order as they are not linked in a sequence.

Use adjectives to describe the animals, such as 'black' or 'dark red' hair.

Gorilla

The gorillas' habitat is forest areas of Africa. They are the largest of the primates, which are the group of animals that include monkeys, apes and humans.

Use subheadings to divide up the text.

Description

Gorillas have black or dark red hair. The adult male gorilla is called a silverback because his hair goes grey along his back.

Pick out an interesting piece of information and place it in a box outside the main text.

Lifespan
Gorillas live up to 30 years.
Food
Gorillas are herbivores. They eat:
o   fruit
o   leaves
o   stems of plants

Illustrate your text with photos and labelled drawings.

Write a caption underneath each illustration.

The gorilla is an intelligent animal.

25

PROJECT WRITING: ANCIENT GREECE

Suppose that a group of you has been asked to write a project on an aspect of Ancient Greece. Before choosing your topic, research all you can about Ancient Greece – from inventors, thinkers, writers and poets to architecture, the city-state and daily life.

Which area interests you the most?

Get ideas through a mindmapping session, and make a list of sections to include, such as farming, gods, festivals, medicine, myths. Organize your ideas into a spidergram, with a topic area, such as 'homes' or 'food' at the centre. Write down key words next to each section.

figs and grapes

fish and seafood

wheat and barley

Food

goat

milk and cheese

bread

olives

FOR SALE

Fine mud-brick and plaster period dwelling with tiled roof – in excellent condition.

- Spacious andron for entertaining friends
- Bathroom with new terracotta tub
- Large courtyard with well
- Mosaic floors throughout

Central Athens. Close to Agora, gymnasium and Temple of Apollo. Early viewing highly recommended! Offers over 5,000 drachmas

Which style?

When you've got enough information on each section, think about the different ways in which you could present it. For example:
- Newspaper report from *The Sparta Daily News* on the invention of the water clock
- Interview with Alexander the Great
- Advert for an Ancient Greek home
- Balanced report on whether girls should go to school
- Sports report on the Ancient Olympic Games
- A letter from a Greek child to their best friend

Daily life in Ancient Greece

The houses in Ancient Greek times usually consisted of two or three rooms built around a courtyard. A wealthy home might have two courtyards and an upstairs floor. The courtyard was the most important part of the home. That is where the family gathered to talk, entertain friends and listen to stories.

Children in Ancient Greece played with lots of different toys. Favourite toys were rattles, wooden hoops, toy animals and dolls made from clay, yo-yos and a horse on wheels that was pulled along. Children also enjoyed outdoor games such as juggling and playing on seesaws.

The Ancient Greeks loved to tell and listen to stories. They told many stories about their gods and goddesses and how the world was created. The most famous storytellers were the blind poet Homer and the slave Aesop, whose fables are still read today.

Ancient Greeks ate lots of vegetables and fresh fruit such as grapes and figs. They made bread from wheat and barley and kept goats for their milk, which they made cheese with. If they lived near the coast they ate fish and seafood. They did not eat meat.

Clothes were made from linen and wool. They consisted of a piece of cloth wrapped around the body and held together with pins and brooches. They wore sandals or went barefoot.

ACTIVITY

Turn this page of information into a report. How can you set it out so that it is easy to read, informative and eye catching?

Remember to :
- use subheadings
- put some text in boxes
- use bullet points and labels
- illustrate your writing with drawings or diagrams
- caption each illustration

SUMMING UP

There are lots of ways to write and present non-fiction texts. They include news reports, balanced reports, recounts and project writing.

Checklist

Here is a list of things to remember when writing non-fiction.

• Audience/purpose

Before you start, you should have some idea who you are writing for. Is your writing intended for a friend, a teacher or someone you don't know? The language you use will depend on who is going to read it.

• Presentation

The way you present your writing is important and will vary according to what you want to say. If it's a recount of a trip, you need to follow the sequence of events. If you're writing for your class newspaper, you need to start with the most important fact.

• Facts and opinions

Non-fiction is mainly about facts but it can also include opinions. Make sure that you distinguish between the two and don't present opinions as facts.

• Research

Make sure your facts are accurate. **Double-check** information you find on the Internet.

And don't forget to:

Take notes
- Have a mindmapping session to collect ideas.
- Use a spidergram, timeline or other device to structure your argument.

Organize your facts
- Start your writing with a heading.
- Use subheadings when you start a new topic.
- Use bullet points to list facts.

Illustrate your work
Include drawings, diagrams and charts to back up your information.

Get together with your friends to produce a school or class newspaper. It's a good idea to have one or two people in charge of each page. You could include pages on:

- school news
- other news
- sports
- book reviews
- puzzles
- interviews

Springboard 1

Springboard 2

Organize a reading club and ask each member to write a review of a book they've read. Keep the reviews in a folder or post them on your school's website, so that others can read them when they're wanting to choose a book to read.

Springboard 3

Write an information book with friends on a topic that you are all interested in. Working with a partner, each pair writes two pages on a particular aspect of the topic. For example, you might write a book on pop bands with pages on:

- when did pop start?

- survey of favourite pop bands

- balanced debate: is listening to music while doing homework helpful or harmful?

- mums', dads' and teachers' favourite bands

- in-depth look at one band

- the best-ever guitarist / drummer / singer

THE DRUMBEATS

GLOSSARY

Active voice	shows that the subject of a sentence is doing the action
Alliteration	when words begin with the same sound for effect
Article	a piece of non-fiction writing about a topic in a newspaper or magazine
Biased	showing favour to one side of an argument
Bold	type that has thick, heavy lines
Caption	writing that explains an illustration or photo
Chronological order	events set out in the order that they happened
Colloquial	informal expressions and phrases used in conversation
Column	a piece of writing that is set to a narrow width. Newspapers are laid out in columns
Contraction	two words that are shortened, using an apostrophe in place of one or more letters
Controversial	a matter or event that people strongly disagree about
CV	a record that lists your education and achievements
Double-check	to check a fact twice, using a second source of reference
Draft	a first try at writing something
Edit	change a piece of text to improve it
Explanatory clause	part of a sentence that gives extra information
Fiction	a story about made-up characters and situations
Font	the style of type used
Glossary	a list explaining difficult or technical words, which is arranged in alphabetical order and placed near the end of the book
Headline	title of a newspaper article
Impersonal language	written in the third person
Index	an alphabetical list of subjects that can be found in the main text
Metaphor	a phrase which says one thing *is* another but isn't literally true, such as 'Sue's a mouse'. Sue isn't actually a mouse but might be shy or nervous like a mouse
Paragraph	a section of several sentences about a subject
Passive voice	shows what is being done to the subject of a sentence
Plot	the events that determine what a story is about
Positive	emphasizing what is good
Questionnaire	a list of questions to be filled in by people, to gather information
Quotes	writing the exact words that a person has spoken
Recount	retelling of events in the order that they happened
Review	an account of a book, film, DVD, play or piece of music that gives the writer's opinion
Scan	reading a piece of text quickly to find out the key ideas or words
Simile	a phrase that compares one thing to another using the words 'like' or 'as', such as 'Hard as nails'
Spell-check	a computer program that checks and corrects spelling
Statement	sentence that gives facts and/or personal opinion
Subheading	a title that has less importance than a heading
Summary	a brief account giving the main points of a piece of information
Survey	finding out opinions on a particular issue

INDEX

NOTES FOR PARENTS AND TEACHERS

- Explain the purpose of organizational features in an information book – the contents page (says what the book is about), the index (provides a list of words, arranged alphabetically, and gives relevant page numbers) and glossary (explains difficult and/or technical words).

- Encourage your child to ask questions about natural phenomena – such as why the sky is blue, why grass is green and why the days are shorter in winter – and to use information books to find the answers.

- Help your child to write and produce a newsletter, turning daily events at home and at school into news items. Remind them to use newspaper features, such as a headline and byline, and to set their writing in narrow columns. The newsletter could also incorporate a sport's article, adverts, a weather report and even a simple crossword.

- With your child, look at adverts in newspapers and on TV. Check for exaggeration in vocabulary and write a list of words that are often used. You could also put together a scrapbook of headlines from a collection of newspapers and ask your child to point out the ones they find most effective, giving reasons for their choices.

- Together, make up some catchy headlines that use alliteration to grab the reader's attention.

- Develop your child's observational skills by collecting a variety of different leaves and asking them to draw each leaf as accurately as they can and to write a simple caption to accompany each drawing.

- Talk to your child about creatures found in and around the home – for example spiders, ants and birds. Encourage them to research each animal through information books and the Internet, and then to write a fact sheet about each one, giving information on lifespan, habitat and feeding habits.

- Encourage your child to draw an animal in its natural environment and to write a caption with an interesting piece of information about the animal. Point out that the drawing should be as realistic as possible.

- Encourage your child to keep a notebook in which he/she can write down observations about their surroundings. Points of interest could include tree species in the area, the sky and cloud formations, and canals, rivers and lakes. As an extension to this activity, he/she could add observations of the seasons changing.

- Together, find topics for debate that are centred around daily life, for example arguments for and against who should take the dog for a walk, who should be in charge of the washing up, whether video games should be regarded as leisure activities or tools for learning.

- Encourage your child to keep records of special days or outings. Put these records into a folder so that the child has a log of some of the things he/she did in a given year.

- With your child, prepare interview questions addressed to a family member or a friend. Questions could be about a recent holiday, hobbies or their favourite time of the year.

- Together, prepare a list of interview questions intended for the child's favourite author. Questions could relate to the author's own life, the writing process, and a particular book's plot and characters.

- Initiate a discussion with your child about a film or play he/she has seen recently or a book that he/she has just read. What did he/she like? What did he/she dislike? Encourage him/her to give reasons and back up his/her statements.

- Encourage your child to keep a diary of daily events.